The V

Other titles of interest:

Nix, Hill, Williams & Bough, *Land & Estate Management*, 4th Edition.
Helliwell, *Economics of Woodland Management*.
Derounian, *Effective Working with Rural Communities*.

The Valuation of Rural Property

A comprehensive Guide to the Appraisal of Farms and Estates

Second Edition

Peter A. B. Prag
MA FRICS

PACKARD PUBLISHING LIMITED
CHICHESTER

The Valuation of Rural Property

A Comprehensive Guide to the Appraisal of Farms and Estates

Published by Packard Publishing Limited,
Forum House, Stirling Road, Chichester, West Sussex, PO19 7DN.

ISBN 1 85341 130 2 (softback only)

Typeset by Dorwyn Limited, 28 The Fairway, Rowland's Castle, Hampshire PO9 6AQ.
Printed and bound by Downland Reprographics Limited, 1 Kingley Centre, Downs Road, West Stoke, Chichester, West Sussex, PO18 9HJ.

Contents

1. INTRODUCTION 1
Tenure; The Markets – Investment Property – Market Differentials – Investors – Land with Vacant Possession – Taxation; Valuation Principles; Purpose of Valuation; Valuation Practice; Type of Property; Area of Practice.

2. LAND WITH VACANT POSSESSION 13
Market Features; Component Parts – the House – Cottages – the Buildings – the Land – Other Factors – Lotting; the Property as a Whole – Assessment of Land – Houses and Cottages – the Buildings – Woods – Location – Planning Considerations – General Condition – Services & Access; Purchase Costs; Depreciated Replacement Cost.

3. MARKET EVIDENCE 36
Property Advertising; Editorial Comment; Statistics; Interpretation; Methods of Sale.

4. LAND SUBJECT TO STATUTORY TENANCY 51
Yield; the Investment market – Traditional Institutions – Financial Institutions – Private Owners & Investors – Developers – Sitting Tenants; Investment Policies – Scale of Investments – Multiple or Single Units – Type of Lease – Location – Age & Circumstances of Tenant – Land Quality – Land Classification – Type of Cropping – Fixed Equipment – Redundant Buildings – General Condition – Rent – Development & Diversification.

5. INVESTMENT CALCULATIONS 80
Basic Capitalization of Rack Rent – Capitalization of Rack Rent after Management & Purchase Costs – Capitalization of Rack Rent on 'Model' Clauses – Allowing for Rent Review – Alternative Capitalization of Rent Review – Dual Rates for Deferments – Life Interests – Market Changes & Inflation – Sinking Funds – Discounted Cash Flows & Internal Rates of Return.

6. VALUATION OF TENANCIES 95
Rental Valuations – Budgets & Gross Margin Analysis – Market Evidence – Rent Reviews – Amenity & Housing; Valuation of Tenancies; Lettings under the *Agricultural Tenancies Act* 1995.

7. LAND LET ON FARM BUSINESS TENANCY 109

8. QUOTAS, LICENCES AND DESIGNATIONS 117
Quotas – Milk Quotas – Livestock Quotas – Arable Quotas; Licences; Designations – Arable Aid Payments – Set-aside.

9. DEVELOPMENT AND DIVERSIFICATION 128
Planning – Development Value – Valuation of Development Land – Hope Value – Blight – Illegal Development – Agricultural Occupancy – Listed Buildings – Trees & Hedges; Covenants, Rights & Easements; Small Plots & Paddocks; Diversification – Regulations – Viability – Agricultural Diversification – Valuation Approach – Financial Approach.

10 WOODLAND 146
Market Evidence – Comparables; Types of Woodland; Grants & Woodland Policy; Yield Classes; Methods of Valuation – Using Market Comparables – Land & Timber – Measurement of Volume of Timber – Establishing the Age of a Plantation – Commercial Considerations – Property Features – Market Factors; Yields & Prices – Insurance.

11 SPORTING RIGHTS 156
Shooting – Pheasant & Partridge – Grouse – Red Deer Stalking – Lowland Deer – Wildfowling; Fishing – Salmon – Sea Trout – Trout – Coarse Fishing; Hunting; Market Trends.

12 MINERALS 166
Mineral Rights; Deposits; Valuation Factors.

13 COMPULSORY PURCHASE 171

14 INSURANCE VALUATIONS 175

15 CHECK-LIST FOR THE NON-AGRICULTURAL SURVEYOR AND OTHER PROFESSIONALS 178
Small Areas of Land; Larger Areas & Complete Farms or Estates; General.

16 TERMS AND INSTRUCTIONS 182
Instructions – the Name & Brief Description of the Property – the Source of Information on the Property – the Date – the Client – Fees; Terms, Conditions & Caveats – Structural Survey – Contamination – Planning – Legal Interest – Tax & Expenses – Disclosure to Third Parties – The Red Book – Valuation Practice.

APPENDIX 1 Sample Valuation Report 188

Useful References and Further Reading 192
Glossary 194
Index 199

List of Figures

2.1	Example of assessing value by component parts.	22
3.1	A summary of land prices as published in *Farmland Market*.	41
3.2	The categories into which the RICS Index is broken down, as published in *Farmland Market*.	43
3.3	A graphic representation of the RICS Farmland Price Index.	44
3.4	A graphic representation of the average price of vacant possession land in the RICS Farmland Price Index.	45
4.1	Bank base rates, 1982–2002.	56
4.2	Agricultural investment yields, 1983–2001.	57
4.3	Agricultural rents, 1981–2000.	58
4.4	Agricultural investments – let land values – 1983–2001.	59
6.1	Gross Margin Analysis of an arable farm of 400 ha.	99
6.2	Fact on Rents: the annual survey of tenanted land in England 2000 – as published in *Farmland Market*.	101
11.1	Capital values of sporting estates in Scotland, 1982–2002.	164
15.1	A diagrammatic illustration of the basic process of appraising agricultural properties.	180

Preface & Acknowledgements

When I was invited to be a Visiting Fellow at Reading University, I was asked if I could help with some of the teaching on the Rural Land Management course. Included within my curriculum there would be the subject of Agricultural Property Valuations for which I was presumably well suited having had extensive professional experience in that particular field. However, there was one problem, I was informed, in that there was no textbook on the matter to which students could be referred and that everything had therefore to be taught in class. Perhaps, it was added as an afterthought, I might like to write such a book?

In doing so, I have had to encompass the most diverse range of rural property issues and this has only been possible thanks to the help and expertise of many friends within the chartered surveying profession. I am particularly grateful to Steve Humphris of AKC and to Anthony Witham, who even suffered themselves to read through the entire manuscript; to Philip Pocock of Adkins, Bill Ward of Ward Williams and Simon Verdon of Fountain Forestry for their advice on Compulsory Purchase, Minerals and Forestry respectively; to Tom Donald for information on Scotland; to Andrew Rettie and Jamie Illingworth of Strutt & Parker and Nicholas Leeming of Humberts for help on sporting issues; and to Simon Pallett of Dreweatt Neate for acting as an informal sounding board for the RICS.

I am also grateful to Charles Cowap, Simon Keeble and David Lewis of Harper Adams College, and to Mike Turner of Peterhouse, Cambridge, for their detailed comments on the script. My thanks go to those organizations that have kindly provided tables and illustrations: AMC (Appendix 1); Blue Circle Industries (Chapter 12); *Farmland Market* (Figures 3.1, 3.2 and 6.2); Fountain Forestry (Chapter 10); Humberts (Figure 2.1); Paul Quagliana and *Shooting Times* (Chapter 11); and Rebecca Hurley and Jim Ward of Savills for numerous illustrations and statisics.

Peter A. B. Prag

CHAPTER 1

Introduction

Farmland must be one of the more diverse forms of property and seemingly one of the more difficult to value. The market within which it is bought and sold, or let, has a limited turnover and is very localized; and such transactions as do occur are mostly of a private nature. Open market evidence can therefore be hard to find and, furthermore, comparisons are difficult as no two farms are alike. In other property sectors one may easily assess such examples as adjacent retail units in a shopping mall or identical industrial premises on a purpose-built estate, or even a terrace of residential houses all constructed at the same time and to a similar specification.

Two neighbouring farms may look similar on a plan, but on actual inspection they are likely to have different types of houses and buildings which will have a major effect on the overall value. Even land that might appear quite uniform on the surface can comprise different soils and have totally different capabilities.

The valuation of such properties involves a very wide range of factors, many of which are of a specialist nature. It therefore needs a high degree of knowledge, skill, and sometimes even inspiration. This book aims to assist the reader in acquiring some of each of these three essential requirements.

Having recognized such diversity and specialization in the agricultural property market, one must accept that there are in fact two, and now possibly three, separate markets, each of which necessitates a different valuation approach. This derives essentially from the question of tenure and from the attitudes and requirements of the buyers and sellers of land within these distinct sectors.

1.1 Tenure

Traditionally, farmland in Britain was largely owned by landlords and occupied by tenants. At the beginning of the twentieth century as much as 90 per cent of all agricultural property was said to be held in this manner. Traditionally, too, farm tenants had been granted a particular degree of long-term security, especially in England and Wales where until recently all such agricultural occupancy was governed by the *Agricultural Holdings Act* of 1948 which effectively gave tenants the right to remain on their farms for their entire lifetimes. The rationale behind this was essentially that, following the Second World War, food production had become such a national priority that it was thought necessary to set structure under which it would be secured for the long term rather th left to the uncertainties of private contracts. The ensuing legislation determined not only the length of tenure but also controlled the terms of occupation, including rent reviews and maintenance obligations.

That situation was exacerbated in 1976 by the introduction of legislation that gave tenants a possibility of retaining their holdings for two or even three generations. Intended as a means of countering the decline of let farms, this new law had the effect, however, of reducing the number of holdings available to rent. This was because either they were being passed to a successor from the next generation or because landlords were loath to re-let any holdings that did become available due to the possibility of not being able to regain possession for such an extended period. These provisions were then revised in 1984 and consolidated in the *Agricultural Holdings Act* 1986, so that any tenancies created after that date were restricted to just a single succession, although the earlier regulations still remain in force for all previously existing tenancies.

Owners of tenanted farmland were therefore generally committed to an ongoing and relatively inflexible system of tenure. This provided a more restricted financial return than from land with vacant possession and created a different level of values deriving, effectively, from a different type of market. These circumstances, combined also with certain tax considerations, made it unattractive to re-let land when it did become vacant and the balance of tenure changed. By the time this legislation was finally replaced by the *Agricultural Tenancies Act* in 1995 only about 35 per cent of Britain's agricultural land was still being let on statutory tenancies, indicating a huge rise in the area of land being farmed in-hand or managed under certain short-term arrangements. In Scotland, where these matters are regulated by the *Agricultural Holdings (Scotland) Act* 1991, an alternative system of tenure had already been introduced alongside statutory tenancies, by means of Limited Partnerships. Land tenure in Scotland is due now to be further modified under a new Holdings Act.

The *Agricultural Tenancies Act* of 1995 effectively gave freedom of contract for any subsequent agreements for the occupation of farmland in England and Wales, but all other existing tenancies still remained under statutory terms, amounting to a substantial area of around 2.5 million hectares. Such traditional leases do of course come to an end at some stage, offering the landlords a reversion to vacant possession; this feature creates a link between two otherwise distinct instances under the 1986 Act of farms being let to ssignable leases, meaning that the owners cannot : tenancies may come to an end, but such cases are ile, the use of new and shorter-term leases follow- the form of Farm Business Tenancies (FBTs), has ty of a third tier of values. Until such time as the market produces actual evidence it is difficult to be sure where these values may lie, but it is likely to be somewhere between those for vacant and traditionally let land and closer to the former, when the letting is only for a relatively short-term.

The average term of Farm Business Tenancies is proving to be no more than three or four years, which suggests that they are being used mostly for the occupation of peripheral parcels of land rather than complete farm lettings. It also makes it unlikely that any such land would be offered for sale during the relatively short time-span of the lease, with the consequence that there is then little chance of any open market evidence of sales of land let on Farm Business Tenancies. In those rather more rare instances when farms are being let for longer terms such as 10 or 15 years, it will be by institutional landlords like the County Councils or the Duchy of Cornwall. These organizations tend to retain their holdings for the long term, so that it would be rare to find any evidence of such tenanted units being sold in the open market. There are occasions too when farms may be offered for sale and leaseback on a longer period, but this is again unlikely to provide sufficient open market evidence.

The ownership of farmland is almost entirely freehold in England and Wales and fee simple in Scotland, whether vacant or subject to tenancy. Leasehold interests are extremely rare and confined mostly to the ownership of woods that may have been sold to the Forestry Commission on a 999-year term, or to internal estate arrangements sometimes involving residential properties or life interests for particular members of a family. Agricultural tenancies are considered to be a valuable interest that may need to be assessed, generally for tax reasons or under compulsory purchase but also occasionally with an assignment or for company purposes.

1.2 The Markets

1.2.1 Investment Property

The exclusive and long-term nature of statutory tenancies means that let land attracts an entirely different type of buyer than that with vacant possession. The owner of a tenanted farm or estate has limited access to the property and in most cases would not expect to gain possession in the foreseeable future. Potential purchasers of let land would therefore be less concerned with the more practical features, such as the style of the house or layout of the buildings, than buyers of in-hand farms who are likely to consider the property for their own occupation.

Tenanted land tends to be bought either by investors or by the sitting tenant and, whereas the latter may effectively have the same views and requirements as someone purchasing vacant land, the market for let farms is determined largely by investment criteria. The sitting tenant may be in a special position to buy and even be expected to pay a premium for his freehold, but the vendor must know too that the only alternative competition would come from outside investors.

Such investors will consider the net income receivable from rent and assess it as a yield on the capital worth. The level of yield will reflect the measure of 'quality' of the property and this will be based on some fundamental features such as terms of the lease, age and calibre of the tenant, land classification and the state of capital equipment. Vacant possession land may well interest investors too but the market is made by farmers and potential owner-occupiers, and they will be looking at different aspects to assess what a property will be worth to them. In these cases, the nature and location of the land and the style and setting of the house, for example, will have much greater bearing than for landlords looking at tenanted farms. The owner-occupier will also expect to achieve a greater return on his capital when farming the land for himself or under some short-term arrangement than when that return is merely the statutory rent payable to a landlord. This is partly to reflect the greater risk attached to farming in-hand as opposed to the relative security of receiving rent and to the additional return expected from an occupier's working capital.

1.2.2 Market Differentials

The differing yields for these two property types creates a differential in value between vacant and let land. The different approach to what defines a property as being of a high calibre or poor quality can then add to the anomaly. This distinction is illustrated in relatively small areas of bare land which can easily attract a high level of competition when with vacant possession, but which would not interest investors if it were let. The former is likely to command a premium price, whereas the latter may be difficult to sell and would only achieve a more modest figure.

Vacant land has in general always commanded a higher price than the equivalent farm let under statutory terms, with the latter tending conventionally to be worth between 50 and 70 per cent of the former. In obtaining vacant possession, an owner can therefore achieve a substantial capital gain, or marriage value, and this has a number of implications for valuation. The circumstances of the tenant are significant, particularly as to whether he is likely to vacate the farm within a few years. While this may depend solely on his age and family circumstances, it can also give rise to a commercial opportunity for negotiating a voluntary surrender. Such issues have a direct consequence on the potential price payable by the sitting tenant and also on the attitude taken by investment buyers.

As has already been mentioned, vacant land now represents the greatest proportion of agricultural property in the UK and it happens also to be the most frequently traded. This is important for valuation purposes because it provides more evidence for comparable price levels, and particularly so in this market which tends to be very localized due to the whole nature of the industry. The circumstances, and the outcome, of land bought and sold in one part of the country can be very different in another area, even if the properties seem to be of a similar type and description. The investment market for tenanted land has, on the other hand, a more national cohesion and is less dependent on local examples, although there are of course some regional influences.

1.2.3 Investors

This market for tenanted land is produced mainly by long-term investors such as the traditional and financial institutions or private individuals or trust funds. The traditional institutions include the Crown and Church Commissioners and the Oxford and Cambridge Colleges as well as some of the larger charitable trusts, most of whom have owned farmland for centuries and many of whom are still maintaining balanced portfolios of mostly tenanted farms. Some public bodies also own tenanted estates, such as the County Councils in England and Wales and statutory authorities like water companies, although these now tend to be sellers of land rather than buyers. The financial institutions comprise essentially pension funds, insurance companies and some unit trusts all of whom made a significant impact on the investment market in the 1970s and 1980s but have since then also already divested themselves of many of their holdings.

The main institutions hold land primarily for financial reasons of capital growth and income performance. Such performance may come from market movements and rent rises and also from special opportunities such as property development or gaining vacant possession. Private investors, on the other hand, may well have an added motive for buying let land arising from the personal ownership of an estate, such as Inheritance Tax relief, or the enjoyment of sporting rights, or simply the status of being a landlord. In such cases, a different valuation approach is needed, even though the property is still a let investment rather than an in-hand farm for own occupation.

1.2.4 Purchasers of land with vacant possession

Land with vacant possession attracts a more diverse range of potential buyers. This can again include investors, whether institutional or now, more frequently, private individuals, but will also involve local farmers keen to expand or other landowners looking to move. Each of these may have a different reason for wishing to buy a particular property, although competing none the less for the same product and thereby creating a single market price. To make proper allowance for these differing market forces requires a different emphasis in valuation technique for each type of property, even though many of the fundamentals are still common throughout. They are also bound by the same essential principles of valuation, as outlined in Section 1.3 below.

1.2.5 Taxation

Farmland has in the past benefited from capital tax concessions, especially for inheritance and for development and other gains. This has at times created distortions in the market, when purchasers have found it necessary to offer premium prices either to secure a particular property against competition or to encourage owners to sell their land to them before some deadline expired. This pressure arose to some extent from the fact that there is generally only a small number of properties available in the farmland market that would suit such purchasers, whether in terms of location, size or type. Even with a three-year period during which capital gains might gain relief from tax by being ‘rolled over’ into a new investment, potential purchasers could still run out of time and effectively pay out part of the tax saving as a premium. The influence of such sales might then be reflected elsewhere in the market as the vendors found themselves able to pay slightly over the odds in order to find land to replace that which had been sold. Tax issues may not now have the same degree of impact as previously, but the possibility of transactions being carried out under such special circumstances still needs to be noted.

1.3 Valuation Principles

Agricultural property valuations may be required for a number of distinct purposes and be carried out under varying circumstances. Such valuations need to follow a single set of principles in order to have a common currency to define the basis on which such assessments should be made and to seek to reduce the number of anomalies that could otherwise arise.

The reference point for such matters is produced by the Royal Institution of Chartered Surveyors (RICS) in the form of a manual entitled *Statements of Valuation and Appraisal Practice and Guidance Notes* and known as the Red Book (a description which dates still from a time when the Institution produced two valuation manuals distinguished familiarly by the colour of their covers). The purpose of the current manual is essentially to set a norm for property valuation throughout the profession in this country and also to give clarification for clients as to the terms under which their assets are being assessed. It establishes certain essential principles that need to be adhered to in all property valuations and also deals with a number of particular situations such as residential mortgages or insurance. It does not, however, at present refer directly to the many issues that are specific to the farmland sector. The valuer needs therefore to be especially aware of how agricultural property valuations should be conducted so as to satisfy the overall RICS requirements. To do this it is necessary to read the relevant parts of the manual for oneself and it is appropriate here to give only a brief summary of the main issues.

The Red Book gives some clear guidelines as to what a valuation figure is intended to represent and defines the hypothetical market situations under which it would apply. It sets therefore effectively a benchmark of a standard set of circumstances which exclude any special factors that would have a distorting effect on a transaction and which would need to be accounted for separately. The most fundamental of these is a definition of Open Market Value (OMV) which, as the name implies, covers the essential features of a purchase and sale between independent parties under normal market conditions. This deals particularly with matters of timing and marketing and the whole issue of producing a figure at one point in time in a market that may meanwhile be changing, and in which transactions take weeks or months to arrange and then complete.

Under OMV, for example, it is assumed that there had been prior to the date of valuation a reasonable time for marketing the interest, as appropriate to the nature of the property and the state of the market. There are other situations, however, when a valuation has to allow for more specific conditions such as in anticipation of a forced sale or when being used for some loan security purposes. These are covered in the manual through two further definitions: of Estimated Realisation Price (ERP) and Estimated Restricted Realisation Price (ERRP). For ERP, the valuation is on the basis that completion of the sale will take place on a specified date in the future, which will have allowed for what is considered to be a reasonable period for marketing. Under ERRP, however, the assumption is that no such time is allowed for marketing and that the sale price would be what a vendor could expect to get on a disposal involving the minimum of delay. The concept of Realisation Prices depends on rather hypothetical considerations and is now generally being abandoned so that OMVs are to be used even in forced sale situations. There is also within the Red Book a category defined simply as Market Value (MV) which is used for deriving an arms-length figure being privately arranged for two parties and also for capital tax purposes.

There can be situations too involving, for example, specialized buildings where the restricted nature of the market needs to be accounted for and where the appraisal will be made on a basis known as Existing Use Value (EUV) or using a Replacement Cost basis.

Essentially, the valuation is intended to reflect the figure that a vendor could expect to achieve for the property after normal marketing and excluding any special circumstances whether on the part of the vendor or of the purchaser. This last point would discount, for example, any premium that might be paid by a neighbouring owner who might have a special advantage in buying adjoining land, such as to achieve economies of scale or to consolidate his holding. It may be of course that either party to a transaction will want to know the best figure that could be justified under such circumstances even though it would not necessarily be supported in the wider market place. This lies outside the normal definitions of valuation but is now dealt with in the Red Book as a Calculation of Worth.

Having established the general principles that must be applied to the valuation of the different kinds of agricultural property, one must also recognize that valuations themselves will be required for a variety of reasons.

1.4 Purpose of Valuation

Valuations can be required for a wide range of purposes, including for example sales, tax planning, legal settlements and loans. The same principles are generally employed in most of these cases although they may incorporate varying methods of approach and come to certain specified conclusions. The main categories can be summarized as follows.

Open Market Valuation. As the description implies, this is a means of producing a figure as close as possible to what the property would be sold for under ordinary market conditions. It must, however, exclude special factors that may have a real influence on the price paid in a particular practical circumstance but which must be ignored when assessing proper market value.

Compulsory Purchase. This will generally involve only some part of the land that may be required for a particular development scheme and not the farm as a whole. The valuation figure will depend greatly upon the open market value but will also account for extra features, such as severance or loss of economies of scale.

Tax. Property may be transferred privately between parties, whether as a gift or settlement or on inheritance, and a value will need to be established in lieu of an open market transaction. Upon the death of an owner this would be for probate, which may involve adjustments for tax purposes but which will still be based upon a full valuation. It can also be necessary to establish a figure that would have applied at the base date for Capital Gains Tax calculations, which is currently March 1992.

Mortgage. Although mortgage loans will tend to be offered at an amount less than the full worth of the property, the valuation will still be made on a normal basis and generally as an open market value but also sometimes specifically as an estimated realisation price, as mentioned in Section 1.3 above.

Insurance. The insurance of premises against perils such as fire and storm will involve a totally different valuation approach, being concerned with rebuilding costs as opposed to market price.

Calculation of Worth. This is an estimate of net monetary worth of a property to a particular party, taking account of the benefits and costs of ownership, whether current or as a future reversion. It is a concept that has been introduced relatively recently in order to meet certain requirements in the commercial property sector where investors are inclined to assess buildings on the basis of future cash flows. There would therefore not normally be circumstances when Calculation of Worth could be applied to agricultural property.

1.5 Valuation Practice

Although valuations may be required for a number of different purposes, it is likely that they will be defined as having been assessed on a particular basis which will generally be in accordance with the Red Book. Where such definitions do not directly fulfil the requirements of the instructing client, the report may be supplemented with advice or comments on those points. One example of this would be when a property is due to be offered for sale and where a recommended sale price would be set so as to cover the potential interests of special buyers which would, however, have been specifically ignored in the formal valuation.

1.6 Types of Property

In this valuation context, 'agricultural property' may range from parcels of bare land to farms or estates and be either vacant or tenanted. Equipped properties such as the farms and estates will include houses, cottages and buildings which may be in agricultural occupation or converted to some diversified use. The land can also encompass woods and sporting rights and there are farms that have an amenity, environmental or recreational function. Furthermore, farmland is the basic raw material for building development and mineral extraction. The valuation must therefore be able to take all these features into account and allow for the possible influence of the various different market facets that may be involved.

1.7 Area of Practice

Despite the fact that a single policy governs agricultural production throughout the entire European Community, the land markets within the separate countries still retain their individual national characteristics. There are differences too in the laws of ownership and tenure and in the manner in which professional matters such as valuation are conducted. The scope of this book is therefore confined at present to valuation practice in Great Britain. Even within these national boundaries there are important distinctions to be recognized in that Scottish land law and market practices differ from that applying in England and Wales.

CHAPTER 2

Land with Vacant Possession

It is appropriate to consider first the valuation of land with vacant possession. Not only does this represent the majority of land being traded in the country but it also incorporates a wider range of factors which can usefully serve to illustrate the various principles involved. Land with vacant possession is also referred to as being 'in-hand' and will cover all properties on which the owner can be reasonably assured of gaining exclusive occupation either immediately or within a suitably short period of notice.

The most fundamental form of possession is where owners farm the land directly themselves. They may of course use outside services such as contractors for some, or even all, of the farming operations, but these arrangements would not normally imply any right of tenure.

An owner may, however, also come to a longer-term agreement with a contractor, perhaps on a yearly period or on a partnership or share-farming basis. In the present context, this raises the question whether in the case of an arrangement that may run, say, until the end of a year it would still be safe to value the property as if it were with full vacant possession.

To establish which form of tenure applies, one needs to go back to the essential basis of a valuation and look at the matter from the perspective of the theoretical purchasers. There are two questions on which they would probably need to be satisfied; first, would there be any doubt that the farming arrangements could be terminated on the prescribed date and, second, will that termination date allow the purchasers timely entry on to the land or would it inconvenience them when taking over the new property? In a business context, such 'inconvenience' could have measurable financial implications if, for example, the purchaser could not transfer his farming operation directly from one property to the other. That might then be reflected in the price offered and therefore in the valuation itself. This in fact raises a number of potential issues regarding tenure, which are also considered in Chapter 4.

For the present, however, it should just be recognized that if a property is to be valued on the basis of vacant possession, it will be on the understanding that an owner is entitled to free occupation of the whole by the time any hypothetical sale is completed. Any variations from this position will need to be accounted for in the valuation. Some may be readily acceptable and require little adjustment. For example, it is not unusual for agricultural properties that are essentially in-hand to include some minor components that are subject to on-going tenancies, such as residential lettings on cottages or leases on off-lying fields. Although the tenants could in many such cases benefit from statutory protection, this need not detract from the main use and enjoyment of the property as a whole in the context of land ownership today.

2.1 Market Features

The vacant possession market is particularly diverse; the types of property vary as do the requirements and circumstances of both purchasers and vendors. Market price, and therefore value, is made by an interaction between supply and demand: supply which is determined by the willingness of landowners to sell, and the demand from potential purchasers. Much will then depend on the circumstances of each particular case and one needs, therefore, to be able to assess the various features of a property that are likely to influence the attitude of the parties involved under the prevailing market conditions. To do so, one can usefully begin by identifying the various component parts of a property which may comprise a house, cottages, buildings, farmland and woods and other features. These can then be assessed not only for their approximate individual worth but also for the particular influence that each may have on the value of the property as a whole.

2.2 Component Parts

As mentioned above, a farm property will tend to include a number of different features each of which will influence the overall market potential of the property and will also have a theoretical value of its own. Even a block of bare land may have to be assessed as more than just a single entity, if it were for example to comprise areas of differing quality, with one part being good arable land and another permanent pasture. By looking at these component parts, one will not only be identifying the full potential of a property, but also recognizing its particular market appeal or shortcomings. This can furthermore provide an initial means of building up a total value for the whole, although it will not necessarily provide the complete answer.

Due to the diversity of the agricultural market, there will be a different emphasis in the value of the components for each type of property. The main factors that one would normally consider are discussed below and can be used to illustrate the principles involved. Although the land itself will usually represent the greatest element of value and be the more important part of a farm, it may be easier at this stage to take things in a different order, looking first at the houses and buildings, which happens also to be the way in which most sale particulars are presented.

2.2.1 The House

In this exercise, the farmhouse may be assessed as if it were a private house with just a garden and perhaps some outbuildings. In doing so, one gains an advantage from the fact that the residential market offers more evidence from which to establish a figure, although there may often be some agricultural factors that need to be recognized before applying a purely residential value. One example would be the proximity of a commercial farmyard. Many farmhouses appear from the front to be picturesque period properties with enviable views over open countryside, but then back on immediately to a range of agricultural buildings which create noise and smell and may restrict the opportunity for a normal garden. In a purely residential market this could be considered a disadvantage and the market price would be discounted accordingly. A more substantial house, for example, might appeal strongly to wealthier buyers who would not be prepared, however, to accept any compromise. On the other hand, with a working farm the house may be of a more modest type and its position next to the buildings might be seen as a practical advantage.

One should not therefore aim to be too precise in establishing these theoretical residential component values and, as so often happens in matters of valuation, it will ultimately be a matter of judgement. In most cases there will be little need for much adjustment but where the house might really stand alone as a private residence, this would be accounted for in the process of lotting, which is dealt with in Section 2.2.6.

2.2.2 Cottages

Until relatively recently, when increasing mechanization began to reduce the need for a large workforce, many farms in Britain employed a large number of staff. Being situated often in remote areas at a time when most workers did not have means of transport and worked what are now described as 'unsociable hours', accommodation was provided on the farm in the form of cottages. Many of these are no longer required for this purpose and are effectively surplus to the agricultural requirements. It is therefore legitimate to assess them for their residential value and indeed many of them might have been sold away from the farm already. As with the farmhouse itself there is usually ready market evidence for country cottages, although again some discount may need to be made for those that may be too close to the farmyard or which are of a limited standard, as may still often be the case. Some may be currently let, either on a long-term statutory lease or now, more frequently, on a shorthold tenancy as established by the *Housing Act* 1988 and latterly under the *Housing Act* 1996. In either case, there is a straightforward means of establishing individual values, using the methods explained in Chapters 4 and 5.

2.2.3 The Buildings

Buildings may be a large feature on an agricultural property and of crucial importance to the farm business, but there is only rarely an open market for them from which to derive an assumed value as in the case of a house and cottages. To incorporate an appropriate figure for them requires a more intricate approach. The actual worth of the buildings to the hypothetical purchaser, and therefore to the farm as an entity, may be measured by the financial advantage that they bring to the business. In other words, what additional costs or loss of income would that business incur if it did not have the benefit of those buildings? This principle may be illustrated by the relatively simple example of a grain store.

Farmers who are able to keep their corn can gain better prices for it by choosing their moment to sell throughout the months after harvest. Without storage facilities, their options are limited to the following:

1. Selling the crop at harvest time ‘off the combine’;
2. Using an off-farm store, such as a co-operative;
3. Building a new store on the farm.

In the first case, one would be selling at the lowest point in the market and without any effective bargaining power. The second situation creates an opportunity to negotiate a better contract, although there will have been some related costs in using the storage facilities. The third option would of course enable one to trade the crop to full advantage, but only after covering the cost of funding the construction of the new building.

Taking this business analysis further, one can see that a hypothetical purchaser would reckon to be able to justify, and finance, a higher price for a farm that had the means of providing a more profitable return on his capital than for a property that lacked such facilities. This principle may well be fundamental to many cases, but again, it will rarely provide the sole answer and one still needs also to step back and look at the market view as a whole.

Rising construction costs and the removal of grant aid, coupled previously with high rates of interest and declining farm incomes, have all meant that new buildings have rarely been viable. On the same basis, a farm that lacked a crucial facility like a grain store might not be discounted to the extent of the cost of providing that facility. There are in fact now only a few occasions when it would be directly relevant to incorporate within a valuation the actual cost of constructing new buildings.

1. Where the farm is effectively prevented from carrying out its business, as in the case of a dairy unit with substandard pollution controls and requiring a new slurry handling system. In that instance, the value when compared to a properly equipped farm would be directly influenced by the full current cost of construction, plus possibly also an additional element to compensate for any inconvenience and uncertainty and for the delay in taking over.
2. In the case of some intensive enterprises using specialist buildings on a relatively small land area. There are two approaches that might be used here.

First, taking the current cost of construction for rebuilding the facilities as existing and then deducting an allowance for depreciation due to age and obsolescence. Known as the Depreciated Replacement Cost (DRC) basis of valuation, this method is intended primarily for the appraisal of premises for which there is no ordinary market and, whilst this does not apply strictly to such situations as pig or poultry units, it can be helpful in establishing a figure within the component-part process of checking an agricultural valuation. An example of a DRC calculation is given at the end of this chapter (Section 2.5).

Second, where the buildings are already of older construction and likely therefore to have only a limited further use, a purchaser may estimate the current cost of constructing the modern equivalent of the units and then discount it according to the number of years remaining before the replacement would be required. The assessment of farm buildings now involves increasingly the question of whether they meet modern requirements such as Crop Assurance or environmental regulations. Even then more modern structures which may appear to be of an adequate design may yet need further investment in order to be fully usable under current market conditions.

Buildings that have been converted to a non-agricultural use may also need to be considered as a separate item. For example, a barn now let as business premises could be assessed on the basis of its investment income, taking the present rent and applying an appropriate Years' Purchase. The basic principles used for this are as for valuing tenanted farms and are outlined in Chapter 5.

2.2.4 The Land

When people talk about farmland values, they often refer to a price per acre or hectare, implying an average measure that can be applied readily to different properties. It would certainly seem easier just to compare evidence of sales of bare land rather than of fully equipped farms which so often have contrasting features and for which there is only limited market evidence. There is however a hidden difficulty in applying such figures to the component calculation of a complete farm.

Bare land will tend to be bought by someone who already has a farm. It would be unusual to buy such land and then equip it with a house and buildings, although this does sometimes occur in exceptional circumstances. Existing owners or tenants may well have special reasons for buying the land. If it can be usefully added to the present farm, they will expect to benefit from economies of scale by being able to spread their current overheads over a larger area. They could then justify a higher price per acre than if the land were being bought as a separate enterprise. Since there is often more than one person in such a position, as there can be a number of neighbours adjoining the land being sold, competition may develop between them and the price will be inflated accordingly. Even without the financial justification of economies of scale, a farmer may bid beyond what would seem reasonable to another purchaser just in order to secure some land that is close to an existing holding and which represents 'a chance in a lifetime'.

Farmland is markedly different to other investments, such as stocks and shares, that can be bought and sold at any time and which have no specifics as to location. Agricultural properties change hands relatively rarely and for them to be of interest to an existing farmer or landowner, they will have to come up within a few miles of their present holdings. One can see that when such an opportunity arises, it will be pursued with a special determination by those within the immediate vicinity, especially at a time when farmers are in funds after a profitable season and when interest rates may be low. This has in recent times been exacerbated also by demand for land to replace areas lost to set-aside or to fulfil extensification requirements, both of which are considered in more detail later. It follows, however, that bare land prices too may not be directly appropriate as comparable evidence for complete farms.

2.2.5 Other Factors

There may then be other features that can be identified as contributing specifically to the value of the whole. These will depend of course on the nature of the property itself and at this stage it is important just to recognize that they need to be identified. This will include such items as woods, sporting rights, diversification and other non-agricultural features. There is also the question of eligibility for grants and quotas and other designations, as explained in Chapter 8. Of particular significance to current property valuations will be entitlement to Aid Payments under the Integrated Administration and Control System (IACS) and the ownership of milk quota which is referred to further in Chapter 8.

A practical example of how this method of assessing the component parts might be applied to a rural estate is illustrated by the table in Figure 2.1 which was published originally by Humberts in *Farmers Weekly* (24 February 1995) and has been subsequently updated.

2.2.6 Lotting

The concept of Component Parts as described in the preceding sections is essentially a theoretical method for building up an approximate value of a farm or estate, but there can be occasions too when in the sale of a property certain areas will be offered separately. An example of such 'lotting' might be a small grass field which when left within the farm as a whole would be priced within the overall figure of £4,200 per hectare, (using the illustration of Figure 2.1). However, if it were feasible to sell this bit of land as a paddock to a local houseowner this could well be increased to £10,000 per hectare or more. A valuer will need to have identified such potential and made proper allowance for it, even if in practice there were to be no intention of splitting off parts of the land in this way.

2.3 The Property as a Whole

Identifying the component parts cannot produce the total answer but is more by way of providing a check-list. It has already been made clear that some of the assessments used in deriving such a figure may have been rather theoretical and might lead to distortions if applied too directly. Also, there may be circumstances when the entire property has a rarity or special attraction that means that it will command a premium over and above the sum of the parts. One needs therefore to consider them in the context of the property as a whole.

It is an interesting feature of the UK agricultural property market that farm sale particulars often begin with a detailed description of the house and then of the buildings followed, almost as a postscript, by just a few lines on the land itself. The land is however frequently the most crucial and valuable feature of the farm and probably the most difficult to analyse. There are furthermore only very limited means by which it might be changed or improved, such as through drainage, whereas houses and buildings can be more readily altered or enhanced if required. In a detailed valuation exercise it is therefore appropriate to begin with the land. Every agricultural property must comprise some land, although not all will have the other components such as house or buildings.

2.3.1 Assessment of Land

Looking again at agents' sale particulars, they seem to use three methods of defining land: by reference to the geological series or to the Land Classification or just by general description. Each of these can play a part in valuation too.

General description is derived from a direct inspection of the land and requires at least a basic knowledge of soil types. 'Medium bodied loam' is an old favourite, if perhaps rather minimal, example of this. The other two definitions are taken from a series of maps covering the country in some detail, firstly the Land Classification series and secondly the Soil

Figure 2.1 Example of assessing value by component parts

Area (hectares)	Item	Value £
10	Principal house, gardens and grounds.	1,700,000
0.2	The Lodge. Former service occupant aged 70 with life tenancy.	100,000
0.2	Keepers Cottage. Service occupant.	120,000
0.2	No.1 Farm Cottage. Agricultural service occupant.	95,000
0.2	No. 2 Farm Cottage. Rent Act tenant aged 50, no dependants.	65,000
0.2	No. 3 Farm Cottage. Assured Shorthold tenant. Let on 6 month term at full market rent	135,000
0.2	No. 4 Farm Cottage. Subject to agricultural occupancy restriction.	95,000
1	Home Farm buildings. Mixed dairy and livestock.	75,000
20	Parkland @ £3,500 per hectare.	70,000
95	Grade 3 pasture @ £4,200 per hectare.	399,000
80	Grade 3 arable land @ £4,800 per hectare.	384,000
30	Woodland @ £2,200 per hectare.	66,000
2.8	Fishing Lake.	20,000
1	Waste areas (roads, tracks, etc.)	0
241	Freehold Total	3,324,000
	500,000 litres milk quota, 4% BF, 80% used.	85,000
	Overall Total	3,409,000

All properties are with vacant possession except where otherwise stated.
Assumed vacant possession value of all cottages is £150,000.
Assumed clean quota price is 17p per litre.

Source: Humberts.

Survey. The former was produced by the Ministry of Agriculture, Fisheries and Food (MAFF – the forerunner of the Department for Environment, Food and Rural Affairs) for England and Wales and by the Macaulay Institute for Scotland and is described in more detail in Section 4.3.6.1 within the context of valuing land let on tenancies, as it has been more widely used in that investment market than for land with vacant possession. The latter is published by the Soil Survey of England and Wales and by the Ordnance Survey in Scotland and maps out to a scale of 1:63,360 (or 1 inch to the mile) the distribution of the various soil series throughout the country. Whilst this cannot provide detailed answers for every field, it does provide a useful point of reference whereby the land can be described as comprising particular soil types. Some of these can be of a localized nature and will be unfamiliar, but there are others such as, for example, the Hanslope and Denchworth series that are found more widely and will be recognized and understood even by the non-specialist.

Although essential to any farm appraisal, the identification and description of the land can present a number of difficulties. First, most of us who may be required to value agricultural property are likely to have only limited knowledge of soil types. Second, on inspecting a farm one might well find the land covered by crops or grass and, third, in so many cases, the soil will vary across the farm. It is quite possible that land in a far corner of a farm, such as in a valley or beside a wood, will be quite different to that which one may see near to the farmstead or by the roadside.

These last two difficulties can be relatively easily overcome, even if it may require extra time and effort. One needs anyway always to have viewed the whole farm, but this will not necessarily mean having to walk every inch of it. Much will depend on the layout of the land and how visible it is from accessible vantage points and whether one can pick out the crucial clues. When undertaking this it is worth remembering that one is engaged in a valuation exercise and not in a detailed survey. One needs only to establish the features that are likely to influence market value; those aspects, for example, that could give cause to the hypothetical purchaser to lower his expectations of farming profit and thereby to lower his price or, alternatively of course, to recognize some advantage and raise his estimation accordingly. None the less, one cannot be caught out either by having missed some relevant detail that should have been accounted for, even if not of major consequence in itself, but where its omission could cast doubt on the credibility of the valuation report as a whole.

Whether or not the land is covered by a crop, a thorough inspection is likely to involve taking samples of the soil both at the surface and immediately below. This is done either by using a soil auger and drawing out a small plug of earth or simply by digging with a spade (discreetly, so as not to offend the farmer!). How useful this will be to the valuer will depend on his ability to recognize the samples and understand their implication. A crucial factor that can often arise is the depth of the top soil itself. A clear example of this may be found in the fens where the surface may show up as a high quality peat or silt, but where only a few inches below lies a bed of clay or chalk. This could have a direct impact on its current farming capabilities as well as its future potential. To make matters worse, it is quite possible that this phenomenon will occur only in parts of a field, towards a boundary or under 'hills' as such variations are rather dramatically known in the dead flat landscape of the fens! Even in such instances, where land prices are among the highest in the country and any shortcomings would have a serious impact on value, one cannot be expected to have physically identified every variation of the soils. It is more a matter of an awareness of such possibilities and of being able to recognize the warning signs or clues.

The crops themselves, whilst hiding the surface of the land can also provide such clues. Rough permanent pasture, for example, would suggest that the land is too wet or heavy to be ploughed. If, on the other hand, a field is planted with potatoes, it would suggest that the soil has to be of a light enough nature to allow for practical harvesting and that it is unlikely to be clay. This is a useful indicator, but it will not always provide the whole answer. The potato land might be prone to drought and dependent on irrigation or have been over-cropped and now infected with eelworm. The rough pasture could also be only a sign that it has never been underdrained and that it might then have an arable potential. Even within a single field variations in a growing crop can indicate something about the land beneath, such as an area that may be waterlogged. Woods that are interspersed among farmland can provide a further clue, in that over the years trees may have been left standing on ground of poor quality that had proved too difficult to cultivate. These factors will be considered in more detail later, but serve for the present to show that a valuer needs also to be able to think like a detective.

2.3.2 Houses and Cottages

The appraisal of houses and cottages may be a more familiar exercise to most surveyors, involving a record of the main features and accommodation and a consideration of its location and condition. How these then affect the value of the farm as a whole has already been considered within the assessment of component parts (Section 2.2) and will be reviewed again later in the context of the valuation as a whole. At this stage it is the more ordinary practical details that need to be identified, to allow a fair comparison with the residential market and also to quantify the issues that a purchaser would be expected to take into account. For example, although a valuation would not normally incorporate a structural survey, some allowance must be made for the apparent condition of a property, particularly if it seems likely that repairs and renovations will be required. As with the land, one needs to have recognized all those features that are likely to be of significant influence on a hypothetical purchaser and these will range from the mundane, such as the type of drainage, to the esoteric as to whether the house is listed.

Establishing the values of comparable residential property in the area can be done by screening advertisements in the local newspapers or by making enquiries off estate agents. By looking at details of properties for sale with the same size of accommodation and in the same kind of location and then finding out the guide price being quoted, one can gain at least a general benchmark of values in the vicinity.

2.3.3 The Buildings

A similar process is required for the farm buildings, although this will need more of an awareness of the agricultural issues than in the case of the house and cottages. For example, the yard may include some traditional buildings in poor repair where a surveyor's first instinct would be to allow for future expenditure to bring them back into good order, just as one would if the main house were in need of maintenance and modernization. However, in the case of the once charming timber-built barns, the modern farmer would have no current purpose for them and it would be inappropriate to discount the value of the whole with an allowance for the cost of repairs as there would be no commercial reason to renovate them. Ultimately this may still need to be considered in a wider market context and whether these buildings might enhance the property's attraction for certain potential buyers, for conversion perhaps to stables or some diversification scheme. One must also be aware of potential constraints, particularly in this context if the buildings were listed and the owner thereby compelled to keep them in sound order. Again, this is a matter that will be reconsidered at a later stage and for the present one must just have registered the main practical implications that the buildings have for the farming business.

These could well involve a wide range of technical features, but the essential principles may be illustrated by a few basic examples. The age and condition of the buildings will have a general effect on the overall value according to the anticipated maintenance expenditure that is likely to be incurred during the foreseeable future and also the timing and cost of their ultimate replacement. Then more specifically, the usefulness and potential shortcomings of each facility has to be assessed.

In, for example, the case of a grain store the following questions arise. Does it have sufficient capacity to hold all the harvest, even in a bumper year? Does it meet current Crop Assurance standards? Do the handling and storage arrangements enable one to hold different varieties in separate sections? What is the capacity of the drier and how convenient is it to run? Is there adequate space for modern grain lorries to manoeuvre safely and be easily loaded? In many cases, grain stores will have been built some years ago at a time when cereals were profitable and when even grant aid was available. Since then, cropping patterns will have changed, yields will have increased and additional land may have been acquired. Meanwhile combine harvesters have a quicker throughput and lorries will also have become bigger. A building constructed to the best standards ten years ago will still look impressive, but it may well now have limitations that affect the performance of the farm and would be expensive to rectify.

The same approach will apply to other types of buildings, according to their purpose and to the specifics of that part of the farming enterprise. Potato stores will need modern insulation; implement sheds must have adequate eave heights and doors to accommodate modern machinery; and chemical stores and fuel tanks must fulfil current safety requirements.

On dairy farms too, there are similar issues. Friesian cows have become bigger over the years and some cubicle houses and collecting yards no longer offer adequate space. Herd sizes will have increased too putting pressure on the workings of the parlour and feed systems and frequently on slurry handling, which is now subject to strict regulatory scrutiny. Milk collection will not always be on a daily basis as before so that the bulk tanks and dairy housing may need to be enlarged.

There will be many other instances which cannot all be listed here. The principle however should be clear, namely that a valuer will have to be able to assess the buildings not just on their structural merits but in the context too of the current and anticipated farming operations. If deficiencies have been identified, allowance may have to be made for the cost of rectifying them or of upgrading the facilities generally.

2.3.4 Woods

It is unlikely that farm woods will need to be valued as commercial forestry. In most cases they will be in the form of spinneys and shelter belts whose significance will be more as an enhancement to the farm as a whole whether for amenity or shooting. This will be considered later in the context of overall market value, but at this stage one would be looking at whether the woods do provide some advantage to the farm, such as shelter for livestock or for the homestead itself or by making the property more attractive to a wider range of potential purchasers. If the farm is of a size and layout to include the possibility of a pheasant shoot, then the woods can be assessed for their particular purpose of providing cover and drives, and thereby contributing to the total style and worth of the property as a whole. More often, however, woods remain on a farm because the land beneath them has never been suited to cultivation and represents therefore effectively an unproductive area.

On the other hand, commercial forestry does not only comprise large areas of conifer plantations in upland areas, and there is a regular investment market in relatively small blocks of often predominantly amenity woods. It may be then that an otherwise agricultural property will include a wooded area with this kind of potential and the basic principles of forestry valuation are therefore dealt with briefly in Chapter 10.

2.3.5 Location

Location is an important factor to most forms of property and has a particular relevance to farmland.

Certain areas gain a reputation for good, or bad, land. Whilst the quality of the soil should have been fully accounted for in the appraisal of the land, this local reputation could still bring an added influence to the attitude of prospective purchasers and therefore to the market value. This may perhaps be more noticeable at the opposite extremes of the market, either for exceptionally good land such as in certain specific areas of the Fens, or where land has become notorious for being particularly poor such as a badly drained sandy heath.

When considered essentially as a business asset by, for example, farmers wishing to expand their operations, then land that lies within a successful agricultural area could have an enhanced value due to the assumed strength of competition in that location, especially if it is easily accessible. In most cases, expansion is only feasible within a limited distance of the original farm, due to the time, expense and difficulty of moving machinery and livestock. If land happens to become available within a convenient distance, then farmers can be expected to bid more strongly than if it were slightly further away. It is a case perhaps of success breeding success. In an area such as Cheshire, land has tended to command higher prices than in comparable dairying districts elsewhere, due largely to the strong local agricultural economy and to the particular locational constraints of expanding a milking enterprise. This is, however, likely to be more apparent for smaller properties than for those of a larger size that do not divide readily among neighbouring owners and which may attract interest from elsewhere in the country. There are also now many areas where the regional agricultural economy has weakened so that local farmers will be less able to buy in adjoining land. This happens often to be in situations where there would be little outside interest either, such as in the uplands, with the result that market values may be seriously undermined. The valuer needs to take account of such regional factors, whilst discounting, however, the special interest of any immediate neighbours whose position is likely to be excluded from the normal open market definition.

It is appropriate in this context to distinguish between what would constitute a purchaser of this kind who has what the Red Book would term a special interest and a very similar situation which might arise however from genuine local market pressure. The definition of Open Market Value seeks to avoid any distortion due to the interest of any buyers whose circumstances enable them to pay a premium for a particular property. This would occur under circumstances already mentioned in Section 2.2.4 when, for instance, a homeowner wishes to buy an adjacent field in order to protect the view from his house, or when sitting tenants are faced with a once-in-a-lifetime opportunity to buy the freeholds of their farms. These would constitute bids that should be ignored when deriving a Red Book open market valuation. It is not uncommon, however, for one property to attract strong interest from a number of potential buyers who might all have a special interest, due to the fact that they are neighbouring landowners and therefore able to benefit from economies of scale by acquiring the additional area, or who might also be swayed by emotion in securing adjacent land. Their bids are likely to exceed those of parties from further afield who would have to consider the proposition purely on its own merits. Yet they reflect a number of diverse and competing interests, and so their position would be seen as being part of an identifiable market which, although creating a premium price, would none the less be sustained even if one or other of the potential buyers were not present. This is in contrast to the earlier case of a buyer with a special interest, without whom the price would be likely to lose its premium.

There are, however, occasions when such special interests should be incorporated within a valuation, or at least referred to, as a supplement to a straight Red Book assessment. A recommendation for sale may well need to recognize this potential premium from local buyers and some appraisals for tax or compulsory purchase will also have to reflect the individual circumstances of the case.

Location can also be a matter of fashion, as with many other forms of property, particularly of course with residential and amenity estates. This influence however would be generally less significant than in the equivalent house market due to the wider number of features in most agricultural premises and to the smaller choice of properties to consider. None the less, there are certain areas, such as parts of Hampshire, where exceptional demand can be generated by an element of fashion and have a recognizable effect on value.

There are some other regions of Britain where farms are eligible for extra grant payments due to their location. While these are mostly in regions of relative hardship, such as the hill lands, the ability to receive those subsidies can underwrite what might otherwise be a lower value. This is mentioned further in the section on Designated Areas in Section 8.3.

2.3.6 Planning Considerations

One's attention has been focused so far on the particulars of the farmland market and the specific features of agricultural property. Land is, however, the raw material for development and can with the benefit of the right planning consents command far higher values than when used just for farming. These implications are considered in more detail in Chapter 9.

2.3.7 General Condition

The condition of the land and buildings will inevitably have an effect on value. When it is evident that some expenditure is required in order to bring the property into a good state of repair, then it is to be expected that any hypothetical purchaser would need to budget for this outlay and deduct it from the price that he would have been prepared to pay had it all been in sound order. Most valuation assessments will have been made on a basis of market evidence which assumes the properties to be in a fair condition, and allowance will therefore need to be made for any identifiable shortcomings. Whether or not such an allowance will match exactly the cost of remedying the fault will depend on the circumstances. If an arable field were to need underdraining, then the commercial price to be paid for it could well be reduced by the estimated cost of such work. On the other hand, where a farmhouse required external repairs and decorations, the likely expenditure on that might easily be accepted as being an unavoidable initial maintenance cost. The valuer therefore need not always have to quantify the cost of remedial works, but he should certainly be able to identify what might need to be done and then determine what effect, if any, that could have on price.

In some cases, it will be a matter of commercial calculation while in others it may simply be a market attitude in that certain types of property might attract higher bids if they appear to be attractive and well maintained. One should be aware too of situations where sub-standard facilities may incur more than just the cost of repairs. Leakages of slurry, silage effluent or of motor fuel, for example, can lead to prosecution and claims for damages if originating from clamps or storage tanks found to be in contravention of the 1991 *Control of Pollution Regulations*.

2.3.8 Services and Access

As agricultural properties may be situated in remote areas they can depend on special arrangements for services such as water and electricity and for rights of access. There are occasions when these will have legal or financial implications and will need therefore to be considered within a valuation appraisal.

2.3.8.1 Access

A farm that is approached by a long private road or driveway may be in an attractive and remote location, but there can be a number of financial and legal issues that affect its value. First, there will be a cost in maintaining the road not only for the farmer's own use but also for deliveries and collection that are often made by large vehicles. A farm road therefore may also have to be widened or strengthened to accommodate such vehicles. Second, such a road might often be shared with other users, such as neighbouring farms or houses. There could then be an obligation to maintain it to their satisfaction or otherwise to cooperate with them on such work. Alternatively, the farm access may depend on a right of way across a neighbour's property which can involve legal or practical problems. Such arrangements are mostly of an acceptable nature and of little consequence on the value of the whole, but there are occasions when they raise legal uncertainties or impending costs which have to be taken into account.

2.3.8.2 Water supplies

The water supply may be shared with neighbours or come from a private source such as the original main estate. Even when it comes from a public main, there can be a long run of pipe for which the property owner will be responsible rather than the water company. There is potential then for disruption due to a failure in the supply or for costs in carrying out extensive repairs or replacements; if these are likely to be substantial they will need to be accounted for in the overall assessment of the property. In the case of private water supplies, it may be worth checking the terms of the agreements for paying annual charges and for contributing to maintenance costs. Also of importance is the reliability of the source in the future at a time when natural water resources are reducing and when health and environmental regulations are becoming more stringent.

On livestock farms in particular, water is of course needed not just to the house and buildings but also to troughs across the fields, some of which may again be at some distance away. Private water supplies to houses now come under increasing regulation as to purity and suitability for human consumption, and therefore can be a risk factor to consider in the overall valuation as to what the costs would be for meeting more stringent requirements or connecting to a mains. Private boreholes are also used for irrigation purposes, generally under licence, and the implications of these facilities are considered in Section 8.2.

2.3.8.3 Drainage

It is not uncommon in rural areas for domestic drainage to be by private systems rather than by connection to a public sewer and there will normally be no valuation implications arising from this arrangement other than when it is shown to be faulty and requiring extensive reconstruction. The drainage from farm buildings, however, can be more serious, particularly in the case of dairying and other livestock units. Landowners and occupiers can now face huge penalties if adjacent water courses are polluted by seepage from farm disposal systems and if those systems are inadequate or in poor repair, the cost of carrying out the necessary remedies will inevitably be reflected in a current valuation. There can also be cases where such costs would be disproportionate to the profitability of the farm, which would then have a major effect on value. For example, if a small dairy farm situated on the banks of a river needed to upgrade its slurry system involving a greater cost than the business could sustain, it might be decided to dispose of the dairy herd and run an alternative livestock venture such as beef and sheep. There would however be a significant differential in the viability and then too in the market value between such a farm when equipped for dairying and when restricted to just beef and sheep.

In some low-lying areas such as the fens in East Anglia, farmland may be subject to Drainage Rates which are levied for the purpose of funding local pumping stations. Whilst these are unlikely to have a significant impact on on the freehold value of vacant properties, it is as well for the valuer to have noted their existence.

2.3.8.4 Electricity

Electricity is almost always provided from the national grid rather than from a private supply but can again be taken over relatively long distances before reaching a particular farm or set of buildings. While this does not represent any undue cost to the consumer once installed, it can have valuation implications where the existing supply is inadequate for modern use. An example of this would be a farm with an outdated grain-drying system that may be connected only to a single phase electricity supply. If the valuation appraisal assumes that a purchaser would budget for upgrading the drying facility, then an allowance would also have to be made for the cost of bringing in a three phase supply. Where this then proves to be prohibitive due to the distance that it would have to be brought, the whole potential of the farm might have to be reconsidered.

There are increasing instances of 'wind farms' being developed within the countryside, often with an involvement by the landowner. These vary in scale from providing a private supply just to the farm to being able to sell electricity into the national grid. There is however as yet little evidence of such properties being bought or sold and no conventional guidance therefore on the way in which the construction of wind turbines might influence the value of the land. Meanwhile, these should perhaps best be treated in a similar manner to on-farm diversification schemes (Chapter 9) with an assessment first of the future reliability of the income generated by the development and second of any disadvantages that might also occur from it, such as a loss of residential or amenity value.

2.4 Purchase Costs

As market value is assessed according to what a potential purchaser would pay for the property, one might consider whether any allowance should be made for the not inconsiderable costs that can be incurred in making the acquisition. Whilst such costs as stamp duty and professional fees may be accounted for in some investment calculations, as outlined in Section 5 (ii), they do not normally form part of a bid price for a property being bought for own occupation and they are not therefore included specifically in the valuation figure. There are occasions, however, when it is known that certain additional costs will have to be met following the purchase and when the buyer's offer, and therefore also the valuation, might be reduced by an equivalent amount. This follows the same principle as for the example of anticipated repairs mentioned in Section 2.3.7 above, but the instances under which it might arise may be less easy to identify. These may be from covenants and undertakings to the vendors or neighbouring owners or even to the farm employees under the TUPE regulations [*Transfer of Undertakings (Protection of Employees) Regulations 1981*].

2.5 Example of Depreciated Replacement Cost calculation

It was mentioned in Section 2.2.3 that the DRC method of valuation was rarely used for agricultural property but that it might be appropriate to a specialized unit and that it would be useful to give a theoretical example. Not only does this illustrate the method that would be used, but it can also highlight a number of more general valuation points.

Intensive livestock units often comprise such specialized buildings that are located on a relatively small site and for which there may not be much evidence of any comparable properties being bought or sold. The DRC approach seeks to establish the sum of the value of the land on which the buildings are situated plus the current cost of construction less a discount for depreciation and possibly also for obsolescence.

The hypothetical example of a poultry farm of three hectares with agriculturally restricted bungalow and five broiler units of 1000 m^2 each would involve the following basic considerations:

		£	£
	Site value: 3 hectares @ £8,300 p.ha., say		25,000
plus	Bungalow	125,000	
less	Discount for agricultural tie, say 25%	31,250	
			93,750
	Current building cost: 5 x 1000m² @ £100 per m²	500,000	
less	Depreciation Allowance, say 25%	125,000	
		375,000	
less	Obsolescence Allowance, say 15%	75,000	
			300,000
	Total Value		418,750

The site value at slightly over £8,000 per hectare is taken as being a little above average agricultural values. This is due partly to the relatively small total area and partly also in recognition of the increasing difficulty of finding sites upon which planning permission would be granted for an intensive livestock unit of this kind. Such planning constraints can have a double effect. First, they create an element of scarcity for any properties for which a valid consent exists. Second, anyone contemplating applying for planning permission on a new site would have to budget for the cost of making the application, in the form particularly of professional fees and of the time taken. Given the opportunity of buying an established site, a potential purchaser would effectively save the cost of those fees and might then be able to justify adding that amount into a bid for the completed property.

The question of an agricultural occupancy tie is addressed in Section 9.1.5. It is not unusual for a small intensive farm unit to include a house that was erected subject to a planning restriction allowing it to be used only by someone engaged in agriculture. In this context, it would be assumed that the bungalow would always be required for use with the poultry buildings, so that the tie would create less of a discount than in other more general cases as described in the later Section. None the less, that does presuppose that the poultry unit will continue to function for the full economic life of the bungalow and also that it would not be unduly discounted when offered as security for loan purposes. As a balance, therefore, in this instance a reduction of 20 per cent was considered appropriate to the overall situation.

For valuation purposes of this kind, building costs may be taken on an average unit basis from a general source such as those mentioned at the end of Chapter 14 on Insurance. Just as in the case of insurance calculations, regard must be had to any particular features of the site that might make construction work more expensive than normal.

Livestock unit costs are sometimes quoted in terms of the number of stock that they can accommodate rather than their actual linear dimensions. In such cases, it is possible to refer to, for example, the *Farm Management Pocketbook* by John Nix to find the average area that each head of stock requires within the building, and then to multiply it up to give an approximate total on which to base one's building estimates.

Depreciation tends to be assessed on a straight line basis over the number of years for which it is assumed that the building will continue to be structurally viable. In the example above, if the buildings were all erected 5 years ago, a deduction of 25 per cent would imply that when new they were expected to stand for about 20 years. Obsolescence is a more difficult concept to identify, although it can easily occur in the case of livestock buildings. Animal husbandry techniques and welfare considerations are both constantly changing so that some buildings that were reckoned to be designed and constructed to good modern standards when they were first erected might no longer satisfy more recent requirements. In the context of poultry, this could currently arise over matters of construction materials or ventilation systems or the size and design of battery cages.

In conclusion, it should be reiterated that for rural property the DRC method of valuation is likely to be used only in specialized circumstances and then cross-checked wherever possible against such market evidence as may be available.

Check-list – Land with Vacant Possession:

- Check that the property is with vacant possession throughout.
- Identify the component features and assess their approximate individual worth.
- Consider the property in its entirety and how the components may enhance the overall value.
- Allow for financial implications of services and state of repair and any legal obligations.
- Assess the prevailing market forces, in general and specifically, for land of that particular type and locality.

CHAPTER 3

Market Evidence

It is clear now that in valuing an agricultural property one must take account of the many different individual features that will influence its total worth. To begin with, however, one needs to be able to establish a general market level for land of an appropriate type and within a similar location. This would ideally be derived from evidence of sales that have taken place, but there are two immediate problems to face: first, that relatively little land comes on to the open market each year and, second, that only a limited proportion of those sales will be on a basis where the actual price is made publicly known.

3.1 Property Advertising

The number of agricultural properties being sold may be limited, but there is an advantage in Britain that many of these will be widely publicized. In other countries in Europe, for example, farmland will tend to be sold more discreetly, without advertising and probably to someone with local connections. In this country, on the other hand, there is a tradition of offering land more openly, so that one can begin to find market evidence by seeing at least what is being advertised for sale. The source of such advertisements will be local newspapers and the weekly farming press, notably *Farmers Weekly* which has national coverage, *Scottish Farmer* for Scotland and *Farmers Guardian* focused particularly on the north-west of England and Wales. Smaller parcels of bare land will tend to be mentioned just in the local papers, with larger properties or those with a degree of wider appeal, such as residential farms, appearing in the farming journals. These bigger or more appealing farms may also be advertised in *Country Life* and *The Field*, as would the more substantial estates and sporting or amenity properties.

Many property announcements are also of course to be found on the Internet, either on sites for the individual agency firms or through a specialist server such as *Property Finder* and *Primelocation.*

From those advertising pages one can gain a record at any rate of agricultural properties being offered for sale. In some cases an asking price will be quoted but otherwise a guide figure can be obtained by application to the selling agents. Two things then need to be considered: how the advertised property compares or contrasts with the land being valued, and the accuracy or realism of the guide price. One can ask as much as one likes for a property, but that does not mean that it will necessarily be sold for that figure. Whether the final price achieved is ever disclosed will depend on the circumstances of the sale. At an auction the matter is by its very nature made public and may well even be reported on in the local press and recorded in the main farming papers. Whilst providing precise evidence of actual prices paid, auction results will not always be a fair comparison to use, as is mentioned in the discussion on methods of sale in Section 3.5. For the majority of sales conducted by private negotiation or even by tender, the final figure will probably only be known to the principal parties actually involved so that a valuer looking for market evidence will have to make his own interpretation on how close it may have been to the quoted guide price.

3.2 Editorial Comment

The media, however, will provide some further clues than just the advertised asking price because the agricultural press does also provide regular commentaries on the property market. Journalists report on what they have learnt about individual sales and may also quote from interviews with the selling agents. This can be by reference to the guide price, as to whether it was at around that level or significantly above it, or may refer to any special circumstances, such as competition from overseas buyers. One can thereby gain a more complete picture of the terms of sale that are reported in this way, although one needs to exercise a little caution when using such information to arrive at general conclusions. Agents have an instinct to publicize sales that have gone well and which therefore not only reflect well upon their abilities but will also give a positive feel to the market and encourage other people to buy or to sell. It is understandably rare for a journalist to be given the same degree of information about a property that has sold badly. Although there is usually no intention to misinform, one must be aware that media comment may tend to focus on the more successful sales and therefore give a potentially inflated view of the market.

One should also be aware that property transactions can take a long time to complete and that these reports will be made only after all the formalities have been concluded. They could therefore be referring to terms that were agreed many weeks previously under conditions that may no longer fully apply, particularly in a volatile or rapidly moving market. When such evidence is still from the same selling season, it will probably only be necessary to make a relatively nominal allowance for any change in attitude that may have occurred in the meantime. There can be occasions, however, when one has to depend on market reports from longer ago and this requires a more quantified approach which is outlined in Section 3.4. This may arise from the fact that the market for vacant agricultural properties tends to be seasonal, for reasons mentioned in Section 3.5 below, while valuations can of course be required at any time and may arise at a period of the year when there is little activity in the market and therefore little current evidence either. It has also already been mentioned that there is, even under normal circumstances, only a small turnover of farmland in the open market and it would be as well now to review the range of data that may be available to the valuer.

3.3 Statistics

When one considers the nature of the land market, one can see that statistical data may well be limited, imprecise or out of date. The most comprehensive source of this information is *Farmland Market*, which is a journal published twice yearly specifically for this purpose. It records not only its own survey but also the results of others that will have been compiled by different methods.

The *Farmland Market* survey is drawn from returns sent in from firms of agents around the country and is unique in that it identifies individual properties and quotes wherever possible the actual price paid. Frequently, of course, the transactions will have been the outcome of a private treaty negotiation and details of the final terms will therefore be withheld and only the price guide given. To that extent, it equates to the type of information that a valuer may already have gleaned from press advertising and commentary, as mentioned previously. However, the survey may include properties that might not have been advertised, nationally or even locally, and does also have the advantage of bringing together a national biannual summary in one source rather than checking retrospectively the different advertising media. Defining each property by a name and address does furthermore give a more precise form of information from which to make one's comparisons than when working from past advertisements or press comments.

Price guides are often quoted by selling agents as being in the region of a specified figure or sometimes that offers are being sought in excess of that figure. This tends then to form the basis of how the outcome of an individual sale will be reported, whether in the press or in answer to an enquiry or in *Farmland Market*. When a property is described as having been sold for a figure 'in the region of' the asking price, one must assume that the sale price was reasonably close to the guide price. Where the sale exceeded the original expectations, it might be quoted as 'in excess of' or 'substantially in excess of' the guide. On the other hand if the outcome was disappointing and the property achieved a figure below the asking price, one can understand that the agent will be disinclined to publicize this apparent failure. It is rare therefore to find evidence of sales quoted as being at less than the guide price, for reasons already mentioned in Section 3.2.

The Department for Environment, Food and Rural Affairs (DEFRA) currently produces statistics on farm incomes and rents, with the latter being published in *Farmland Market*. The Agricultural Land Price Index that was previously issued by MAFF through its advisory agency, ADAS, is now produced by the RICS and also quoted in *Farmland Market*, although published separately too on a quarterly basis. This index draws again upon results submitted by agents throughout England and Wales but is then compiled so that it records average price levels on a national and regional basis and according to the different agricultural sectors. By not referring to specific properties it has the advantage that it can use actual sale results rather than guide prices, but to preserve confidentiality these can then only be produced in the form of average figures. A copy of the summary of these figures as published in *Farmland Market* is shown in Figure 3.1.

The Inland Revenue also produces data on land prices which was published by MAFF (now DEFRA) and by the Valuation Office as part of the biannual Property Market Report. This covers the whole of England, with a similar service also being provided by the National Assembly for Wales, Agricultural Department, and the Scottish Department of Agriculture, but again referring to general averages rather than specific instances. Although more comprehensive than other series, in that the Revenue should have access to details of every land transaction that takes place whether publicly or privately, there are some disadvantages in using it for current valuations. First, there tends to be a time-lag because the statistics are taken as at the time when duty is paid, which can be many months after the transaction was agreed or had taken place. Also they will include sales or transfers made outside the open market, such as within families or between trusts.

A number of agents produce statistical commentary on land prices on a national or regional basis and drawn either from transactions and

CALP/RICS FARMLAND PRICE INDEX 1992-2002
CURRENT AGRICULTURAL LAND PRICES
VACANT POSSESSION: ENGLAND AND WALES

see notes: 5 & 6	Sales in 3 mths ended	No./Reported Sales	Area Sold 000 ha	Average Price £/ha	Weighted Average Price £/ha	Price Index 1985 = 100
	Mar-93	74	3.5	3425	3731	84
	Jun-93	81	3.1	3432	3228	73
	Sep-93	58	2.7	3998	4139	93
	Dec-93	48	1.9	3497		
Eng only	Mar-94	15	0.83	3426		
*	Jun-94	24	0.64	5594		
*	Sep-94	27	1.6	5327		**1995 = 100**
*	Dec-94	26	0.8	6219		**Price Index**
	Mar-95	73	3.9	5160	5187	100
	Jun-95	117	5.2	5349	5336	103
	Sep-95	163	11.1	5469	5672	110
	Dec-95	91	4.8	5269	5556	107
	Mar-96	114	4.8	5737	6023	116
	Jun-96	141	6.9	7275	6731	130
	Sep-96	313	15.7	7011	7019	135
	Dec-96	225	9.7	7092	7046	136
	Mar-97	118	6.5	7093	6906	133
	Jun-97	208	9.1	7160	7263	140
	Sep-97	210	8.8	7758	7870	152
	Dec-97	183	8.5	7171	7166	138
	Mar-98	113	5.2	7364	7494	145
	Jun-98	133	5.5	7557	7545	146
	Sep-98	164	7.1	6595	6411	124
	Dec-98	149	7.9	6974	6812	131
	Mar-99	84	3.6	6123	6376	123
	Jun-99	138	4.5	7002	7337	141
	Sep-99	152	9	7440	7186	139
	Dec-99	151	6.9	7009	7204	139
	Mar-00	90	3.9	7472	7660	148
	Jun-00	120	4.9	6840	7211	139
	Sep-00	134	7.2	7101	7082	137
	Dec-00	128	5.9	6171	6461	125
	Mar-01	73	4.5	7109	7273	140
	Jun-01	70	4.3	7504	7967	154
	Sep-01	137	7.5	7,473	7,169	138
	Dec-01	130	5.8	7,925	7,327	141
	Mar-02#	71	3	7,953	7,906	152
	Jun-02#	73	2.8	6,807	7,092	137

Notes:

1 The RICS Farmland Price Index commenced in January 1995. Data is collected on the same basis as that collected for the earlier CALP series, except that the value of milk quota is excluded from RICS prices.

2 The average price is derived by dividing the total value of sales by the total area sold.

3 Since this average price can be much affected by a small number of large or exceptional sales, a weighted price is calculated by weighting the average price for each region x size-group category by the area of land sold in that category during the base period. For the CALP sales reported to September 1993 the base period was 1986-1988. For the RICS Farmland Price Survey sales reported from March 1995 the base period is 1995-1997.

4 The weighted price is expressed as an index relative to the first figure of the base period.

5 Figures for the most recent two quarters(#) are provisional and subject to revision as further information becomes available.

Figure 3.1 Reproduced with kind permission of the Royal Institution of Chartered Surveyors

valuations in which they have been directly involved or from an analysis of publicly advertised sales. *Farmland Market* contains also a section also of regional comment by agents, which can provide background information for valuations although not intended to be precise enough to use as valuation evidence.

3.4 Interpretation

The market evidence tends therefore to be of a general nature, either because no precise sale figures can be given or because the statistics give average trends rather than specific results. These average figures, however, do have some use as a valuation tool, as a benchmark against which to check individual cases and as a means of working in different periods.

There will be occasions when a valuation figure is required for a previous point in time, particularly in tax matters such as capital gains computations or in legal disputes. It is quite likely that the valuer will not have adequate past records for that period, in which case it may be helpful to establish from the published indices the percentage by which average land prices have changed since that time and apply the same rise or fall to the present valuation figure. Conversely, there may be situations when the only comparable market evidence for a particular property is from a year or two ago. This evidence might still be usable if one is able to update it by applying the relevant average percentage change shown in the index.

The same approach may have to be adopted when the lack of immediately comparable evidence means that one must look at recent sales of properties that are either similar to that being valued but in a different locality or, alternatively, may be closer by but of a different type or quality. In both cases one will need to make adjustments to allow for the main differences and, although this will inevitably rely on interpretation and judgement, it can be made easier by analysing the component parts. For example, if the farms are in the same vicinity, but have a very different residential content, then this could be quantified to some extent by considering the theoretical impact that each of the houses might be assumed to have on the total values. If there are similar farms in different locations, it should be possible to determine whether there is a general premium in one of the areas where there has tended to be a premium, as has been the case, for example, for dairy farms in a location such as Cheshire.

Although the published statistical indices can therefore provide an important point of reference for valuations, one should also always be

Farmland Price Index

(Royal Institution of Chartered Surveyors)

England and Wales

Quarterly average of all sales (12 acres and over) excluding land sold for development or forestry, gifts, inheritances and compulsory purchases.

VACANT POSSESSION

Sales in the three months ended	Number of sales	Area sold acres	ha	Average price £ per acre	£ per ha
Sep 01	124	18,327	7417	3068	7582
Dec 01	112	11,532	4667	3196	7897
Different land types	**Number of sales**	**acres**	**ha**	**£ per acre**	**£ per ha**
Bare land Sept only	84	6528	2701	14,002	5793
Bare land Dec only	80	3867	1600	14,449	5978
Over 20ha Mar	68	15,854	6416	3157	7802
Over 20ha Jun	56	9980	4039	3265	8067
Under 20ha Mar	37	875	354	2919	7214
Under 20ha Jun	56	1576	638	2760	6819

TENANTED

Sales in the three months ended	Number of sales	Area sold acres	ha	Average price £ per acre	£ per ha
*Sept 2001	-	-	-	-	-
*Dec 2001	-	-	-	-	-

* Insufficient data

VACANT POSSESSION

Breakdown by region of vacant possession sales for the fourth quarter 2001 (where 3 or more sales recorded) excluding residential and other farm types.

Region	Area sold acres	ha	Average price £ per acre	£ per ha	Number of sales
North West	-	-	-	-	-
Yorkshire and	608	246	3668	9063	7
Humberside	-	-	-	-	-
East Anglia	2199	890	2943	7273	17
West Midlands	423	171	2975	7350	13
South West	2402	972	2885	7128	28
South East	4317	1747	3745	9254	22
East Midlands	593	240	3262	8060	10
Wales	427	173	2149	5311	9
North	-	-	-	-	-

VACANT POSSESSION

Breakdown by farm type of all vacant possession sales received for the fourth quarter 2001.

Farm type	Area sold acres	ha	Average price £ per acre	£ per ha	Number of sales
Residential	2849	1153	6745	16,668	32
Dairy	729	295	3193	7889	6
Arable	6148	2488	3419	8073	52
Mixed	2449	991	3267	8448	12
Other	215	87	7082	17,500	7
Beef/Sheep	2231	903	2503	6185	42

The information provides the basis for a new, regular, up-to-date land price series to replace the discontinued CALP series. Data is collected and analysed quarterly for the RICS by the Centre for Rural Studies at the Royal Agricultural College, Cirencester.

Figure 3.2 Reproduced with permission of *FARMLAND Market*

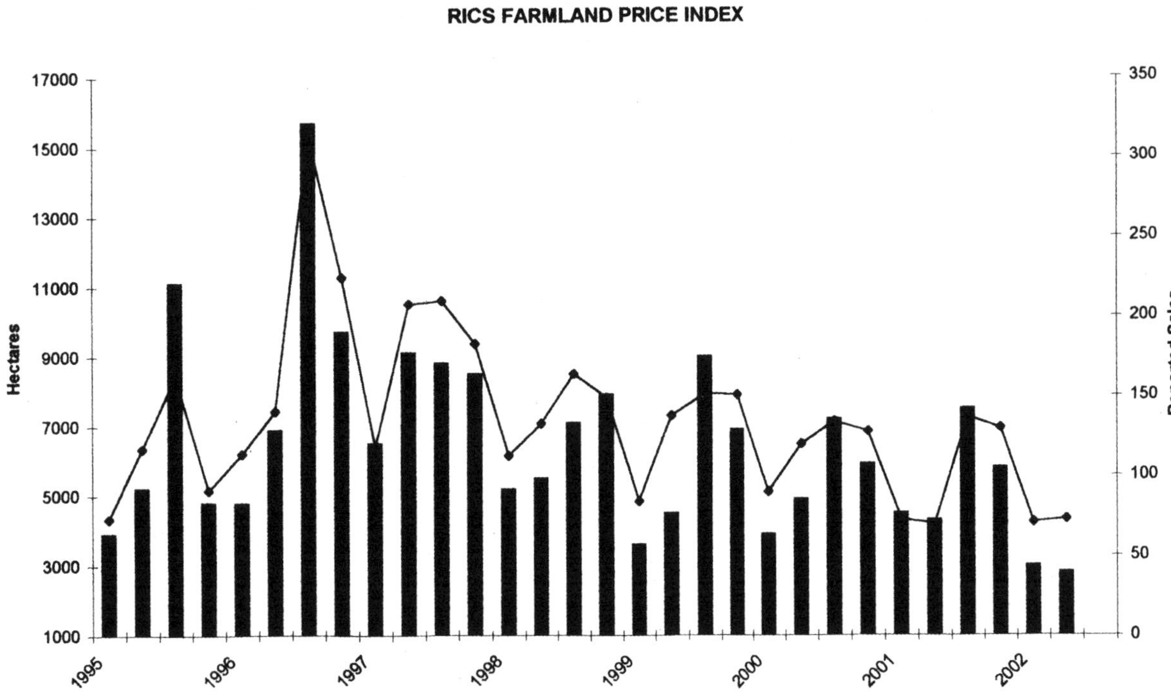

Source: RICS

Figure 3.3 Although dealing sometimes with only a relatively small sample, the RICS index appears to be a good indicator of average market activity in that there is such consistency between the number of sales and the total area, as shown here with the columns representing total area sold and the line being the number of reported sales. Potential distortions are avoided by excluding special situations such as properties of less than 5 hectares, or with a high residential or development content.

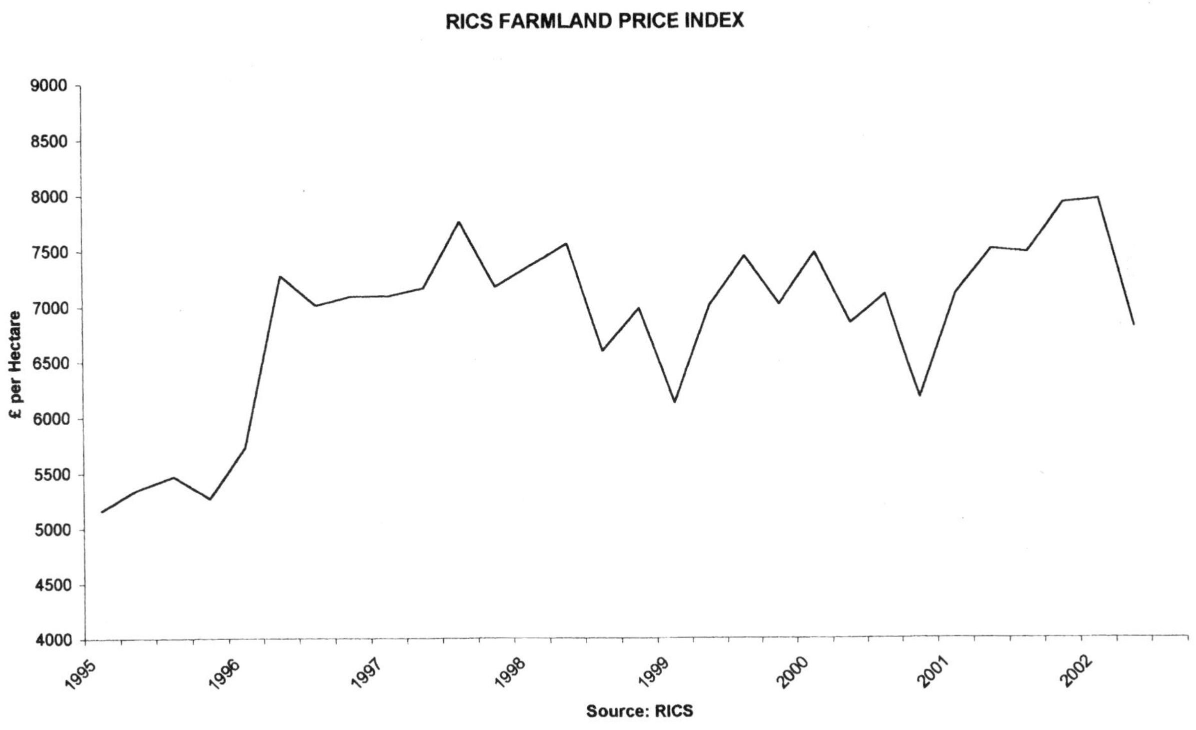

Figure 3.4 The movement in average price seems in some degree to follow that of turnover as shown in Figure 3.3.

aware of their potential limitations. The figure that is mostly widely quoted by farmers, agents and journalists alike is the national average price per acre or hectare. When compared with the equivalent figure for a previous time period this may well give an indication of a market trend, but it will not necessarily be applicable as a direct comparable for any one specific property. It has already become clear that values will vary according to the type and location of the land and the particular attributes of its component parts. Even though some indices are broken down into regions and farm types, there are still too many possible variations for any very precise comparison to be made. The categories into which the RICS index is broken down is illustrated in Figure 3.2 reproduced from *Farmland Market*.

Furthermore, there are times when the average figures have had to be based on relatively small samples so that they are more liable to potential distortion from individual transactions. These occasions of a shortage of transactions can occur just at a time when the market is particularly volatile due to a lack of land for sale and an excess of demand from potential purchasers, and when valuation evidence would therefore be all the more crucial. One might expect that under such circumstances, a small market sample would coincide with a high average price being recorded. Indeed, when looking at the published results for the first three years of the RICS index, as shown in Figures 3.3 and 3.4, it appears that periods of increased market activity have coincided with a direct but modest price rise. It is noticeable, however, also that during this time the number of sales recorded was relatively small at mostly between 100 and 200 per quarter. It is interesting to note too that the number of sales responded closely to the area sold, implying that even in a small market there was little distortion from any large individual transactions. It would be wrong to put too much emphasis on such potential statistical anomalies and, indeed, the RICS index is a unique and much-used tool within the market place. None the less, one should be aware of the fact that there can be times when the farmland indices are based on a very limited sample and that this may hide potential distortions and lead to a misinterpretation of trends.

3.5 Methods of Sale

Having established how best to work within the likely shortage of evidence in the agricultural land market, there is a further consideration to bear in mind when using published data as much could depend on the manner in which the property had been offered for sale.

One needs at first to look at auction sales. These are after all the only source of specific results of transactions, and it is important therefore to know whether they are truly representative of the rest of the market. One hears often enough of auctions when purchasers may have been carried away by the occasion and continued bidding beyond the expectations of the vendors, resulting in record prices being achieved. On the other hand, there can also be instances when the reverse happens and bids fail to materialize and the property has to be sold quietly after the auction for presumably less than had been originally anticipated. Clearly, one must consider the circumstances of such sales before using the results as evidence of average market activity.

A property may be sold either by auction or by private treaty. An auction is normally a public affair, but can also be arranged privately as a means of concluding a sale between a small number of seriously interested parties whose interests cannot be satisfactorily reconciled by normal negotiation. Such private auctions are relatively rare and their results are anyway unlikely to become widely known. Private treaty sales are arranged generally by negotiation between the two parties or their agents, but can also result in the vendor asking potential purchasers to submit best offers. These may be little more than a written version of someone's final offer and still be subject to contract or they can be formal tenders leading to a commitment at least on the part of the purchaser to conclude the transaction on the specified terms. Either way however, there is no reason why those final terms would be made public, although when it is known that a property was sold by such means one may assume that it was because there had been strong interest from potential purchasers which would result in a competitive price being paid.

The auction method appears to have two advantages. First, when a bid is accepted the purchaser will immediately sign a binding contract and pay over a deposit, whereas under private treaty most sales are agreed initially 'subject to contract' leading to several weeks of uncertainty while the formalities are concluded. Second, as an auction is public, it would seem to demonstrate openly that the best possible price had been obtained and that all those involved in the sale – agents, trustees or executors – had fulfilled their tasks correctly. There will be, however, many situations when these factors would not apply, which explains perhaps why auctions are not more widely used in the rural property market, and why one needs to be careful when using the results for valuation evidence.

The advantage to the vendor of securing a binding contract at the time of the auction itself can be achieved only at a cost to the potential purchasers. Anyone bidding at the auction will have to be fully prepared to proceed with the acquisition, and will therefore have to have gone to the expense of implementing surveys and legal searches beforehand without knowing, of course, whether these might not be abortive if the property is sold to another party on the day. This may be acceptable for larger properties involving purchasers with substantial resources but could be a deterrent to potential buyers of smaller lots. There are instances when a big farm may be offered at auction either as a whole or in lots, when the smaller parcels do then attract successful bids, but these will tend to be from neighbouring landowners who would not be dissuaded by the cost of searches. On the whole, however, when vendors opt to sell smaller properties by auction, in order to benefit seemingly from the factor of demonstrating that the best possible price had been obtained, they may in fact be doing the opposite as potential buyers are put off from participating.

For the vendor too there are considerations of scale, as there are greater costs incurred in setting up an auction which may not be justified in these smaller cases. A successful auction depends upon having attracted a sufficient number of potential bidders to the venue on the day to create the necessary degree of competition. This requires a greater degree of advertising prior to the sale than would otherwise be required, and there is then additionally the cost of hiring the auction room itself. Furthermore, as the auction is to be formally binding, the sale brochure will incorporate effectively a legal contract and be of a more expensive format than private treaty particulars. The need to attract a sufficient number of bidders applies of course to all auctions, whether large or small, and this will depend on the nature of the property, the circumstances of the sale and the state of the market. In a weak market particularly, people are already disinclined to buy and would then be unlikely to be drawn into competitive bidding. In an auction held under such circumstances, it becomes only too evident that there is no firm interest in the room and the vendor will be exposed to an unfavourable negotiating position afterwards than might have been avoided with private treaty.

It will normally be both unfeasible and unnecessary for a valuer to try to identify all these influencing factors when looking at auction results as market evidence, but it is nevertheless important to be aware of the possible distortions that could arise.

In Scotland, the legal system allows for a more binding arrangement for agreeing terms of a sale, equivalent to the formal tenders mentioned previously. As a consequence, best offers tend to be used instead of auctions so that there are fewer publicized results of sales. On the other hand, Scotland does have the advantage of a system of land registry known as the Register of Sasines in which the price paid for a property at its last sale is recorded and made available to anyone enquiring for such information. Although not providing full details of transactions, this register can provide information on values of specific comparable properties or on the price at which a property itself was previously sold.

Quoted guide prices are also not always as definitive as one might expect and one needs again to be aware of the background circumstances. A selling agent has three basic options when advising vendors on price: to fix a figure at which one would sell if that amount were offered; to set a lower figure and ask for offers in excess of that amount; or to pitch the price at a high level and be prepared to reduce it if necessary should prospective purchasers not be prepared to offer that much.

The last of these can occur if an agent is keen to win the sale instruction and thinks of impressing the client by suggesting that he can achieve a particularly high price. This may be done as a subterfuge knowing that in all likelihood he will have to explain later that his original expectations cannot be achieved. It can also be quite legitimate under some circumstances, however, such as with an exceptional property or in a rising market, to set a high figure to avoid the possibility of failing to get the best possible price. One must bear in mind too that clients can themselves have fixed, and often inflated, views of the value of their own properties and may not be dissuaded from having the asking price set accordingly.

Asking for offers in excess of what may seem a modest guide price can be a valid tactic with a more difficult property or under unfavourable market conditions. The agent is hoping thereby to attract the interest of a number of potential purchasers and then create a degree of competition between them which might be worked up to a higher price than originally quoted.

There is also always the possibility that the selling agent had not assessed the situation correctly and that the guide price was simply wrong – valuation can, after all, be a difficult business! It can happen too that a property attracts a purchaser in a special position, such as reinvesting capital gains, who may be pressured into paying more than would be normally justified. As a rule, however, one should remember that marketing tactics may be involved in setting a Recommended Sale Price and that this can therefore be different to an Open Market Value.

Mention has been made previously of the seasonal nature of the farmland market and the effect this may have on the available evidence at certain times. Farms have traditionally tended to be offered for sale in the late spring or early summer. This has been due to the fact that land tends to look well at that time of year and that it avoids the busiest period in the arable calendar when prospective purchasers might not be able to find opportunity so easily to view properties. Viewing is also facilitated by the long daylight hours and, furthermore, sales agreed during that season can be completed at the conventional handover date of Michaelmas in the autumn, enabling the buyer to take over at what was traditionally the start of the new farming year. These arrangements are now less rigorously held than in previous times, particularly in the tenanted sector, but there will still be periods in the year when there may be little being advertised in the market and when it would therefore be more difficult to find any contemporary evidence of open market sales.

Check-list – Market Evidence:

- Obtain publicly available data from property advertising and editorial comment.
- Allow for private nature of most rural transactions and for need for interpretation of quoted evidence.
- Use published statistics and indices as indicators of trends.
- Make adjustments, when necessary for differences in time, location and property detail.
- Be aware of the possible significance of the different methods of sale.

CHAPTER 4

Land Subject to Statutory Tenancy

The valuation of land let on an agricultural tenancy requires a different approach to that with vacant possession. This arises primarily due to the different circumstances of the likely purchasers of such property. In the case of land with vacant possession, the market is determined generally by the requirements of buyers who intend to occupy the property themselves and are therefore directly concerned with the kind of practical features that were outlined in Section 2.2. Farms that are subject to tenancies are usually of interest more to investors or to the actual sitting tenants. The tenants are in a special position that will be considered in Section 4.2.5. For valuation purposes one needs firstly to assess such properties from the perspective of the investment market.

The fundamental forms of farm tenancies were mentioned in Section 1.1 as coming under the *Agricultural Holdings Act* 1986 or the *Agricultural Tenancies Act* 1995. Other statutory occupations do also arise in rural property, although usually on minor parts of a farm or estate and of lesser significance to the valuation overall. These include principally residential lettings under the *Housing Act* 1988 (Assured Shorthold Tenancies), or occasionally still under earlier *Rent Acts*, and some business occupancies under the *Landlord and Tenant Acts*. Prior to 1995, extensive use was made of various arrangements for the short-term occupation of land under licence or partnership, for example, but these are now rare and would in any event not normally be considered as requiring a valuation as if subject to tenancy. In Scotland, however, where there has been no equivalent change to that of the 1995 Act, fixed term occupations can still be encountered under partnership arrangements.

It has already been emphasized (in Chapter 1) that traditional agricultural leases have provided long-term security for the tenants and that investors would rarely expect to gain possession in the foreseeable future. To them, the property is essentially a source of rental income, which needs to represent an appropriate return on their capital. Its value to such investors will therefore be determined by monetary rather than practical factors: matters of rent, income growth, investment yield, and prospects for capital growth both from general market movements and from the property's potential. Whilst the assessment of tenanted land will still involve those features considered in the valuation of farms with vacant possession, they will be of a more secondary nature as the emphasis focuses initially on the investment criteria. Although investors may have little knowledge or interest in the practical aspects of farming, the valuer will of course still need to assess these in some detail in order to provide the necessary overall appraisal.

Before looking at these particular issues, it may be useful to consider the concept of **investment yield**. In purely financial terms, investments are made on the basis that they are likely to show a return on capital that is appropriate to that particular set of circumstances. The general level of return is determined by overall economic conditions such as bank rates and inflation. Ultimately after all, an investment will be measured by the current cost of money and the anticipated erosion of capital value over time. Within those parameters, each type of investment will attract a greater or lesser degree of confidence or interest and this will in turn be reflected in the yield that is required from it. Essentially, a holding that offers a secure income that appears likely to be protected against inflation will need to show a lesser return than one which carries a greater degree of risk. When the initial income is the same for both of these alternative examples, such as being in the form of a full open market rent, the difference in quality will be reflected in the price that investors are prepared to pay. Price, or capital value, is calculated by representing the known income as a rate per cent return; income divided by the yield that it is expected to show, produces this capital sum. To illustrate this with a simple example, if a 'good' investment is expected to show a four per cent yield, while that of a lesser quality would need to produce five per cent, on an initial rent, say, of £100, an investor would pay £100 divided by 0.04 or £2500 for the former and only £2000 for the latter. To determine what yield to apply to a particular property requires an assessment of market attitudes, an analysis of the rent and an appraisal of the quality of the farm both in the perception of investors and in its practical ability to produce the required rent. One of the essential elements of value of tenanted farms is therefore that of yield.

4.1 Yield

Investment yield is determined partly by market perception and partly by the quality of the property itself. The latter draws upon many of the same factors as those that one considers for the valuation of farms with vacant possession, albeit with a different emphasis as discussed below. Market perception depends upon a number of different elements, not all of which are related specifically to agriculture. Investors will decide upon a range of yields that they will be prepared to accept for the general sector of agricultural property. This will in turn be determined by the relative attraction of alternative forms of investment and by the outlook for the agricultural and other property markets.

The assessment of the agricultural sector in relation to other investments depends upon a wide range of factors. When interest rates are high and the returns on financial investments are correspondingly high, one might expect that agricultural investment – in order to attract funds – would need also to show higher yields than at times when money rates are low. As value is effectively the price that a hypothetical purchaser would agree to pay for a property at the particular time, it is important to know whether investors would need to see agricultural yields reflecting the rise or fall of these other rates. Such rates can be quite volatile, as illustrated in Figure 4.1 showing changes in money rates over a 20-year period, and so if that were to be the case, then farmland values would also follow a similar fluctuating pattern. Meanwhile however, one can see from Figure 4.2 that agricultural investment yields have been comparatively stable over the same period, due largely to the illiquid nature of the market meaning that investors would not expect to be able to react swiftly to changes as they would in other sectors. Meanwhile too, farm rents have shown a particularly steady trend, as indicated in Figure 4.3, and as value is a function of rental income and investment yield, one can see from Figure 4.4 that tenanted land values appear to have responded almost directly to the perceived change in agricultural yield. One conclusion that may be drawn from this is that there will be times when investors will accept a wider differential between the returns from agriculture and those from other sectors. If one can identify the reasons behind this, one may establish the main determinants of value of tenanted farmland.

The first factor to recognize is that property transactions take time and that prices and yields cannot therefore respond with the same speed as some other markets. The image of a city trader with a telephone clasped to each ear whilst scanning flickering screens has no counterpart in the property world. It can take weeks, rather than seconds, to complete a property deal and it may already have taken as many weeks beforehand even to find the right proposition. Property investment yields cannot therefore move with the same rapidity as those in the financial markets. That in itself, however, may not explain the relative stability of the property sector. When other rates rise significantly above those being required for tenanted farmland, one might expect investors to concentrate exclusively on those opportunities with better yields and not even consider buy such as agricultural property whose return lagged behin sectors. If that were the case, then at any such moment one sumably only sell let land at a yield equivalent to the higher obtained elsewhere which, based on prevailing rents, wou ...ean a corresponding drop in price. To an extent this does happen in that investors may be less keen to buy farmland under such circumstances and certainly there can then be fewer transactions from which to derive one's statistical evidence.

It becomes necessary then to consider more closely the **potential** demand from investors rather than look only for examples of actual deals done. In a relatively small market such as farmland, when there can be little or no evidence of offers being made or accepted, valuations may well have to rely to some degree on an understanding of what positions investors, and indeed owners as the possible vendors, would be prepared to consider even though these are not being put to the test at the time. This is of course a matter of judgement and interpretation and underlies the need for a knowledge of the various factors that influence the investment market.

4.2 The Investment Market

There are five main categories of investors that currently own or have a potential interest in tenanted farmland: traditional institutions, financial institutions, private investors, developers and sitting tenants.

4.2.1 Traditional Institutions

The traditional institutions comprise such bodies as the Crown Estates, the Church Commissioners, the Duchies of Cornwall and Lancaster, some of the Oxford and Cambridge Colleges, and a few old established Trusts. Although their individual circumstances and the extent of their holdings are varied, these organizations are seen as having a common role in the investment market in that they have all owned let land for long periods of time and tend still to take a relatively longer-term view. This implies that they are likely to retain their properties rather than sell and that they would have little cause to be buying in further land. In that case their position would have little influence on the investment market other than being seen as withholding a potential supply of tenanted

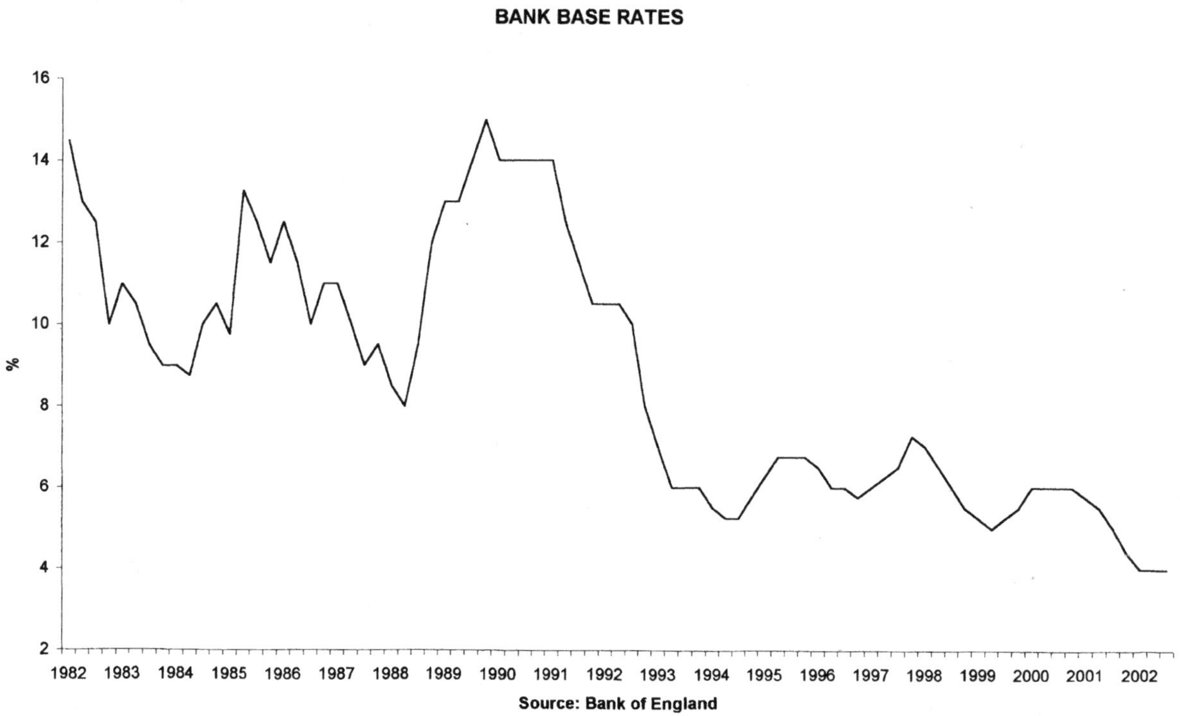

Figure 4.1 Although investors' attitudes to property yields may be influenced by what is happening in other investment markets, the sharp fluctuations in money rates over 20 years as represented by Bank Base Rates were less pronounced in the agricultural sector, as illustrated in Figure 4.2.

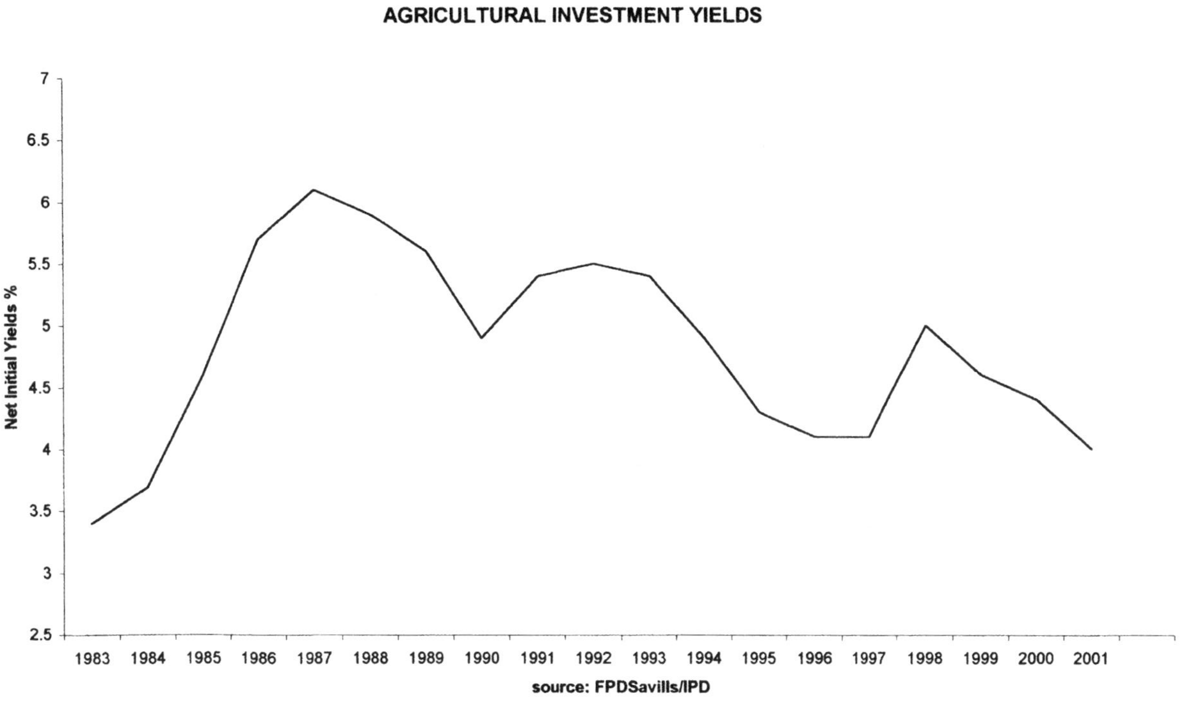

Figure 4.2 Having traditionally shown lower annual returns than any other property sector, agricultural investment yields rose steadily during the 1980s at a time when farm incomes were beginning to decline in real terms due to the Common Agricultural Policy and when alternative propositions were showing stronger short-term growth.

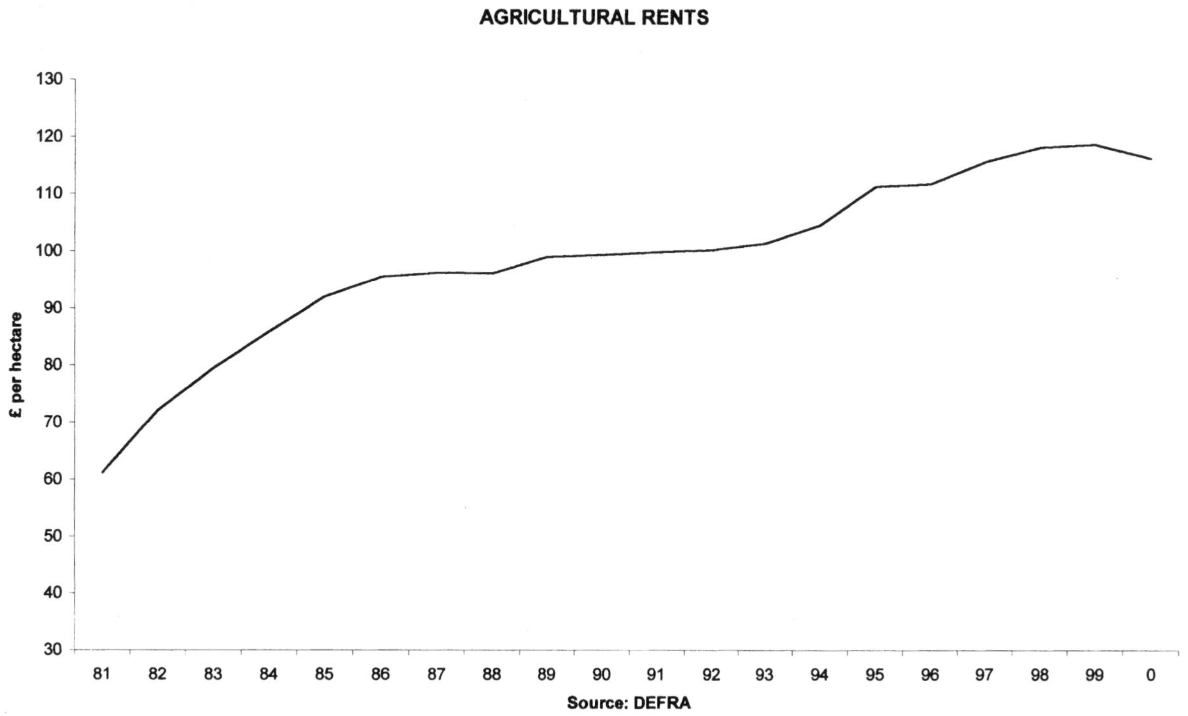

Figure 4.3 Despite falling farm incomes during the 1980s and at the turn of the millenium, rents continued to show modest growth. This reflects the impact of Farm Business Tenancies and that statutory rent reviews may lag behind other economic events by up to three years and may also simply have been witheld rather than negotiated downwards.

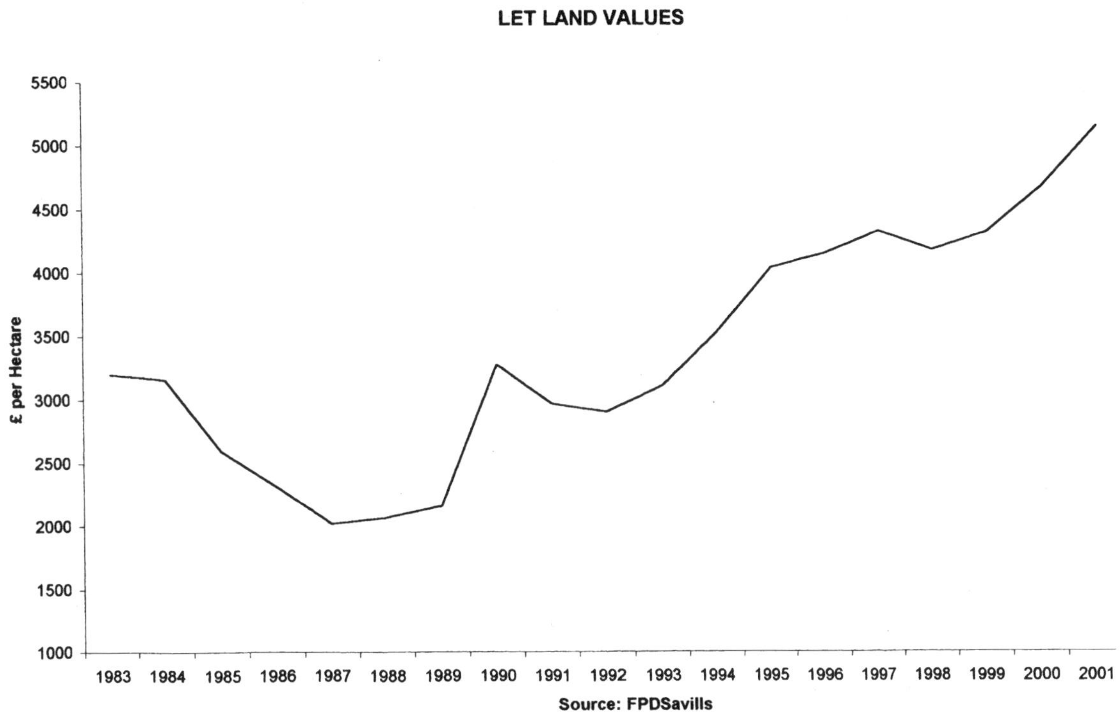

Figure 4.4 That land values seem to mirror in reverse the movement in yields, as shown in Figure 4.2, suggests that the agricultural investment market is determined essentially by net initial returns rather than by reversionary elements.

properties. In fact, these estates are generally of a traditional nature in terms of types of farms and of tenure and this encourages a degree of active portfolio management by the owners, resulting in land being both bought and sold according to circumstances.

4.2.2 Financial Institutions

The financial institutions comprise in the main UK pension funds and insurance companies. Their involvement in the agricultural market dates, in most cases, from the 1970s when they became a major buying force of tenanted land. More recently, however, they have become net sellers of these holdings and their presence in the rural market is now far less evident and focused generally on land with development potential.

4.2.3 Private owners and investors

Private landownership may be either in the form of traditional estates or of property bought more recently for investment purposes. In some ways these equate to the traditional and financial institutions mentioned above, although they do each have different positions in the market. The owner of a traditional estate, such as one of the historic titled families, may well have held the land for a long time and be inclined to take a longer-term view, particularly in those many cases where the property is being held under a discretionary trust. Some will however be more vulnerable to tax and other financial constraints in contrast to their institutional counterparts. This makes it more likely that such land could come up for sale or, alternatively, be taken in hand when a tenant dies or retires. In the former case, there can be a direct influence on the investment market while the latter has the effect of reducing the supply of land let for more than the relatively short term of a Farm Business Tenancy. In general, such traditional landowners have not been significant buyers of land in recent times.

Yet there has been demand for let land from private investors, both from within the UK and from abroad, whether as individuals or in the form of Trusts. Their situations and therefore their reasons for acquiring farmland are very varied, as considered in the section on policy below. They may currently represent only a relatively limited financial force, but their changing presence is none the less an influence on the market.

4.2.4 *Developers*

Property development companies as such are not usually ir tenanted land, but it can happen that funds will be available to purchase farmland on which there may be potential for some alternative use. Such purchases would not be made on the basis of a long-term agricultural investment but on the justification of unlocking development gain or for the potential of securing reversions and resales in the future. Once more this is referred to in the section on policy below.

4.2.5 *Sitting Tenants*

There has been an increasing trend that when let land has come up for sale it may be bought by the sitting tenants. This arises from a number of circumstances which relate largely to the level of demand from other investors, to the profitability of farming and to the cost and availability of finance. These matters are considered further below as is also the question of how the interest of a sitting tenant should be treated for valuation purposes.

4.3 Investment Policies

To derive an open market value for a particular tenanted property one needs therefore to assess how it would be perceived by the various potential buyers. For a sale to be successful, a property will have to satisfy the requirements of investors at the time. If one can identify those requirements and then see how well the property might fulfil them, one can establish from which sector of the market there is likely to be interest and at what level. To do so one has to consider their investment policies at the time, the relevant characteristics of the property and the prevailing state of the market.

The main property features that have a direct bearing on investment attitudes are:

- the scale of investment,
- whether it comprises a single unit or an estate of several farms,
- whether the leases are on traditional or modern terms,
- location,
- age and circumstances of the tenant,

- land quality, probably as indicated by the classification maps,
- type of cropping,
- extent and ownership of the fixed equipment,
- redundant buildings,
- general condition,
- opportunities for development and other diversification.

4.3.1 Scale of Investment

As a general rule, the institutional investors have had little interest in smaller properties as these are less convenient to manage. These big organizations are accustomed to dealing in large financial tranches, often of well over £1,000,000, and so there would need to be special reasons for making an acquisition below this level. There are also likely to be difficulties in the practical management of a smaller farm, even if let on modern terms, unless again there were special circumstances such as when the land is close to existing holdings. One cannot of course give a precise definition as to what range of size each type of investor would consider and much will depend on the state of the market at the time and on the particular circumstances. By and large, however, a farm of less than around 100 hectares or a let holding of bare land will depend mostly on the interest of the sitting tenant or possibly of a private investor, especially one with local connections. Farms of between about 100 and 200 hectares may interest a slightly wider range of private investors and possibly some of the more traditional institutions. Above this level, the market is more open, although sitting tenants may perhaps be less likely to contend for the larger single units while still being a force on the bigger estates that comprise farms of a more modest size. This very general example may illustrate the principle that one needs to identify to whom the property is likely to be of interest and then consider the potential level of demand from that type of investor under the prevailing market conditions.

4.3.2 Multiple or Single Units

It is then also evident that size alone may not mean that the property is bound to interest larger investors. A big estate of several smaller let farms, or a portfolio of a number of holdings scattered around the locality, may not necessarily attract the more commercial investors. This arises mostly from the perception that management may be more difficult and less efficient, and it may also reflect an assumption that smaller farms will be less profitable and more vulnerable to change. None the less, there will be other big investors who will take a different view, recognizing that a more diverse holding can give greater opportunities for capital gain in the future through amalgamations and reversions. How a fragmented or diversified estate of smaller farms is assessed will depend therefore on whether it has this further potential and whether the appropriate type of investor is active in the market at the time.

4.3.3 Type of Lease

The main reasons why investors may buy tenanted rather than vacant farms are that it is assumed they are easier to manage and the income is more secure. The occupier undertakes all the farming business, with all its technical requirements and considerable risks, while the owner's responsibilities might be limited to maintaining and insuring the premises or even to doing no more than just collecting the rent! The landlord's position can be identified by reference to the terms of the lease. Whilst all agricultural agreements prior to 1995 had to follow statutory provisions, without any scope for opting out, there was none the less freedom to come to one's own arrangements as regards repairs and insurance. Many of these are determined according to the 'Model Clauses' laid down originally in 1948 and now covered in England and Wales by SI 1473 under the *Agriculture (Maintenance, Repair and Fixed Equipment) Regulations* 1973, as amended by SI 1988 No. 281.* These clauses have provided an accepted framework whereby liabilities are defined and allocated between the two parties. In a lease incorporating modern terms, the tenants become responsible for all repairs and insurance by virtue of the fact that the agreement requires them to take over all those liabilities otherwise defined as being due to the landlord. Not only is the landlord thereby spared the expense of repairs, but it will involve far less management time and expertise, and therefore fewer professional costs, than with the more traditional statutory clauses.

In this latter case, the landlord will be liable for repairs to all the main structures as well as for insurance. This arrangement dates from the time when tenanted farms would form part of a big estate where the landlord

*The equivalent Scottish legislation was consolidated in the *Agricultural Holdings (Scotland) Act* 1991.

would have maintenance staff and might also have the means for providing home-produced building materials such as timber from the woods and stone from a quarry. In a modern context of course this is rarely the case and a landlord will have to buy in materials, use outside contractors and also possibly find the necessary professional supervision. All this will be expensive and inconvenient. The expense may well be defrayed in the relatively higher rent paid under that form of lease, although the continuing liability for the maintenance particularly of older buildings is difficult to quantify. The inconvenience of having to deal with such matters will be unwelcome to those investors whose approach to agricultural investment is essentially financial and who do not really have the facility for the more practical aspects of ownership. The traditional institutions, however, will be accustomed to such leases and already have in place management systems capable of dealing with them efficiently. Some private investors may also be prepared to take on such commitments, but the demand for properties let on traditional terms will generally be more selective than that for the more modern form of tenure. If demand is more limited then so too may be the price and therefore the value of the property in the open market.

The impact that the different management clauses may have on net investment income and on the yield applied to it and therefore on value is considered in Chapter 5.

The terms of a lease cannot be changed unilaterally, although there may be occasions when this can be done by mutual agreement. It is possible therefore that tenants of farms that are let on traditional terms may take over the repairing liability from their landlords, and the valuation could be enhanced through the improved returns and yield. This might arise as part of some wider agreement in which the landlord provides new equipment or brings in another member of the family as joint tenant, although there could also be a financial advantage to the tenant in taking over these obligations. A landlord will generally have to use contractors to carry out repairs, incurring substantial costs that are ultimately reflected in the rent. Tenants, on the other hand, could use their own or local labour at a lesser cost while also paying the relatively lower rent that would apply to a full repairing and insuring lease. Whilst the valuer should be aware of such possibilities when assessing any foreseeable developments on a farm, these agreements are not necessarily as straightforward as the above example implies and, unless there are special circumstances, the income should still be assessed as net of the traditional outgoings.

Mention has been made in Section 2.3.8.3 of Drainage Rates being charged in some low-lying areas. Under some traditional leases these outgoings are shared between landlord and tenant, in which case they would need to be taken into account when assessing the owner's net income position.

4.3.4 Location

Location will have a bearing on the value of any property, whether vacant or let farmland or also residential or commercial. With agricultural investments, the area in which the land is situated can be of particular significance. Conventionally, most purchasers of tenanted farmland have preferred the larger, more commercial units and these will tend to be found more readily in certain parts of Britain and be unusual or unknown in other districts. Geographically, the west side of Britain is more hilly and with greater rainfall. The majority of farms produce grass rather than arable crops and either are compact, such as a family dairy unit, or comprise extensive, rough hill land. As one goes further east, however, the land becomes more level and the climate less wet and there is a predominance of arable units many of which have developed into relatively large enterprises. There have been situations when investors looking for a bigger and more commercial unit would expect to find it in the eastern part of Britain, and probably also more in the southern half of the country. A large arable farm of good commercial quality may well exist elsewhere, such as in Herefordshire, but the main potential investors will have been less familiar with that area and less likely to have other holdings there. Their interest will therefore be less keen than for a similar proposition in Cambridgeshire or Lincolnshire. To that extent if the farm being valued would still interest these more commercially orientated investors, in terms of size and tenure as well as quality, then there could be a noticeable discount due to its geographic location.

As the larger investors have their preferences for certain types of locations and for bigger holdings, an agent preparing a tenanted property for sale in one of these areas would take particular note of any that already own land there. One would expect them to have a special reason to buy additional land, in order to increase the size of their investment and to improve the economies of their management, and they might therefore be expected to show a greater degree of interest than for an equivalent property in a neutral location. The implications of such special interest and the resulting 'marriage value' has already been discussed in Section 2.3.5.

The possibility of future development or other reversionary gain depends upon location, not so much according to the wider regions of the country but more specifically as to whether the land lies close to an urban area.

4.3.5 Age and Circumstances of Tenant

When land is let on the old statutory terms, the tenants have a right to remain on the holding for as long as they continue to fulfil the terms of their agreement and, in some cases, this security may pass to their successors. Were they not to continue on the holding, the owners would gain vacant possession and benefit from the higher values that such land normally commands. Where this is unlikely to happen in the foreseeable future, the property will be assessed as an investment property according to its current yield and the other factors considered by a non-farming landlord. If however there is a possibility that the tenants might leave the farm in due course, then an allowance will have to be made for the ultimate enhancement in value that this would bring. The methods of calculating such reversionary value are dealt with later in Sections (vi) and (viii) in Chapter 5, but for the present it becomes clear that whether tenants are old or young or whether they have sons or daughters working with them has a direct influence on value. Other factors also have to be considered, such as the tenants' general competence in running the farm and ability, and willingness, to fulfil their obligations. Statutory agricultural leases are almost always in the name of an individual or joint individuals such as a husband and wife or a family partnership, rather than to companies. Assessing the calibre of tenants is therefore more like the residential than the commercial property market, and there is no convenient measure to refer to such a covenant. Much will depend therefore on the individual tenant or tenants, not only as to their abilities but also their overall circumstances such as whether they own or rent other land, or may want to buy the freehold or alternatively to be bought out. A cynic might reckon that a farm let to a bad tenant should in fact be worth more than one occupied by a better farmer: it is more likely that he would fail to fulfil the obligations of his lease and would thereby forfeit possession, leaving the landlord with a significant capital gain. In practice, however, this would be unrealistic as it has been extremely rare in recent times for tenants to be forced to vacate their farms. There are two main causes under which a notice to quit against a bad tenant might succeed: either non-payment of rent or bad husbandry. The latter is difficult to establish and would need to be proven before a tribunal. The former may be easier to establish but because prompt payment of rent is so crucial to a tenant, even those in the hardest financial circumstances will always tend somehow to find the necessary amount on the day.

As the differential between tenanted and vacant values has been so significant, it is important for a valuer to be familiar with the grounds whereby a statutory tenant may be required to quit his holding, however rare it may be for any of them to arise.

In England and Wales, security of tenure under agricultural leases that commenced prior to September 1995 was determined by a series of Acts dating from 1947 and culminating in the *Agricultural Holdings Act* 1986. Essentially, a tenant has the right to remain on the holding for his entire life provided that he fulfils the terms of the agreement. If the landlord were to issue a notice to quit, the tenant could serve a counter notice and the landlord would then need to establish before the Agricultural Land Tribunal that his grounds for seeking possession came within the few cases specified by law, which can be summarized as follows:

- non-payment of rent;
- bad husbandry, meaning that the tenant's management was unacceptably poor;
- development, in that the land was needed for non-agricultural purposes;
- greater hardship, suggesting that the landlord had a more pressing need for the land than the tenant.

In anticipating a reversion to possession one must therefore be confident that a valid case can be made and will also have to allow for the time that may be required for this to take place. Normally a valid notice must be served on or before the term date 12 months prior to when possession would be sought. Even in the case of non-payment of rent or of bad husbandry, minimum statutory periods of two and six months apply. An accurate valuation for investment may have to take this into account as referred to in Chapter 5.

There are circumstances too when a landlord may not be sure of gaining possession even upon the death of the current tenant. In 1976 the *Agriculture (Miscellaneous Provisions) Act* gave the successors to a tenant who had died, or was retiring, the right to apply to take over the lease instead. This was effectively revoked by the 1984 *Agricultural Holdings Act* and any tenancies created since that date up until the introduction of Farm Business Tenancies in 1995 would be for a single lifetime only. The legislation was not however retrospective and the 1976 provisions therefore still apply to all leases entered into before August 1984. When valuing land let on any such older tenancy one may not assume that the owner will benefit from a reversion to full vacant value upon the anticipated death or retirement of the existing tenant without first checking the possibility of a valid claim for succession. The conditions under which such a claim may be upheld were carefully defined in the 1976 Act and then consolidated in the *Agricultural Holdings Act* 1986 and the valuer will need to consider these in detail if there seems to be any likelihood of a claim being made. Essentially, the applicant must demonstrate that he or she is both eligible and suitable. This means that they must be financially and technically competent to run the farm and that a substantial part of their income is derived from the holding.

The 1976 Act allowed for this opportunity to be available even to the next generation behind the original successor. Whilst the intentions of the present tenant's grandchildren can hardly be identified so long in advance, this may yet have implications for a valuation. It can arise that an arrangement may have been agreed in the past between landlord and tenant allowing a younger person to take over the lease, without necessarily qualifying under the Act, but on the understanding that it would be recognized as a first succession so that only one further application might be made in the future.

To establish whether a statutory claim for succession could arise, the valuer needs first to check the date of the original agreement and then to consider not only the age and calibre of the existing tenant but whether there is someone, usually but not necessarily a son, wife or daughter, who is participating in the farm business now and could possibly qualify for succession at such time as the tenant were to die or retire. There are of course difficulties over this as not only is one prejudging a future state of affairs, but succession claims are themselves quite complex and decided by a tribunal. For valuation purposes, however, one needs at least to have established the possibility of a succession claim and then determine whether this might influence the attitude of investors to the risk of not ultimately gaining vacant possession. The way in which such situations are quantified is dealt with in Section 4.4 and Chapter 5 below.

This legislation applies only to England and Wales. In Scotland, where the relevant regulations were consolidated in the *Agricultural Holdings (Scotland) Act* 1991, a tenant has for many years had the right to bequeath his lease to a successor, without any limitation and with little opportunity for counter claim by the landlord. The successor does however have to be a son or daughter, or spouse, whereas in England and Wales the applicant need only be the equivalent of such a member of the family. The investment market for tenanted land in Scotland has therefore already made some allowance for such eventualities and there is less need to make adjustments to valuations. Furthermore, land legislation in Scotland is undergoing radical review which may impinge upon values, especially of tenanted holdings.

4.3.6 Land Quality

The assessment of tenanted land depends upon the same features as those outlined for vacant land in Section 2.3.1 but with an essential different emphasis. Whereas the hypothetical buyer for in-hand land would be assumed to have a direct interest in the practical farming of the land, the investor in let farms has neither the knowledge nor indeed the need to know as much about the details of the soil. Investors tend therefore to put a greater emphasis on one particular measure of land quality that has a lesser significance in the buying and selling of vacant properties, namely land classification.

4.3.6.1 Land Classification

Mention has already been made in Section 2.3.1 of land classification and of the series of maps produced originally by MAFF (now DEFRA) for England and Wales and the Macaulay Institute in Scotland. To appreciate their relevance, or otherwise, to agricultural value one should consider briefly the origins of these maps. They were produced in 1976 for the purpose of providing some guidance to Local Authority planners as to the 'quality' of farmland. At the time, an increasing amount of development was taking place on rural sites and planners were open to criticism for allowing this on land that was reckoned to be of prime agricultural quality and which should therefore have been retained for that purpose in preference to some alternative site which did not have the same productive potential. To help planners recognize which land might be of better quality, the Ministry produced a series of maps covering eventually the whole of England and Wales and defining land in five different grades.

These grades were effectively a measure of versatility. Grade 1 was ascribed to land that was capable of the widest range of cropping, including such special products as salads, onions, bulbs and carrots but including also, if it were ever appropriate, all the more ordinary crops like cereals. Grade 2 has still a high degree of this versatility but would not be suitable for some of the more specialized Grade 1 crops while still including root crops such as sugar beet and potatoes. Grade 3 is that bit more limited and restricted to combinable crops such as cereals and oilseeds or grazing, while Grade 4 is effectively unploughable and best left in grass. Grade 5 is land that is of poor quality and with severe limitations due to its geological type and also perhaps its situation and local climate. Whilst the original maps show the country designated under these five grades plus a sixth category of non-agricultural use, a sub-category of Grade 3a has been introduced subsequently. This came about in response to the fact that Grade 3 covered a relatively wide range of soil types and that it was felt necessary to be able to distinguish the lighter and more workable land from the rest. It has become possible therefore to get such land re-classified as Grade 3a and this description can appear in sale particulars, for example, although the maps have not been reprinted and will still show only Grade 3.

In Scotland, a similar system of classification is used albeit using the different term 'Class' instead of 'Grade'.

In the present context one can see that non-farmers would find this scale of land types a convenient and easily understood means of determining whether a farm was good or bad. Investors, and particularly the big financial institutions, would therefore be inclined to pay more for the higher grade of land because this denoted, to them at any rate, higher quality. In many ways they have been right to do so, not least because rents tend to be higher on the better grade of land. These rents reflect the fact that specialist cropping can be expected to produce higher cash returns than more ordinary farming and that there is furthermore a limited supply of such land. This shortage is exacerbated by the fact that many of these crops cannot be grown on the same field more frequently than in some cases one year in seven and need therefore to be rotated around a wide range of this scarce land. These factors also influence the vacant price of high-grade land, enhancing its value above that of other types and encouraging investors of let farms to pay an equivalent premium. In accepting this they have tended also to accept a lower initial yield from top-grade land, increasing the price still further. This general example illustrates the manner in which land classification has been used in the investment market and has influenced value, but it also raises two conundrums.

First, Grade 1 cropping requires a degree of specialism and carries with it an element of risk. It can happen that, for example, carrots which might normally produce a financial return per hectare of over twice that of Grade 3 arable crops, have on occasions to be ploughed back in when the market becomes oversupplied and prices fall so that they become unsaleable. The higher land price and the lower investment yield may not therefore properly reflect the inherent risk in achieving a consistently high return. Second, the land classification maps, although produced to a relatively large scale of one inch to the mile, or 1:63,360, were never intended to show a detailed profile of every field and have a claimed accuracy to only 80 hectares. It is easily possible that when land is assessed in the more detailed manner described for vacant land it will not reflect the classification ascribed to it. In such cases, an agriculturist will be guided more by the physical evidence and the implications which it may have for the actual farming potential than by the grading. The vacant value will not therefore be enhanced by the high classification to the same extent as when considered as a tenanted investment. The different attitude to land grading therefore creates a potential distortion between the two markets and it explains also why land classification does not have the same significance when dealing with vacant land.

Land classification will therefore influence value through demand from investors, but it cannot be the sole consideration. Even if the owner of let land is not directly concerned with the practicalities of farming, the return on his investment will depend upon the level of rent which will in turn depend to a large degree on the physical features of the land. There are aspects of the land that could affect the farming, such as being heavy and badly drained, and that a tenant could argue a restriction to its profitability and therefore a limit to the rent. To that extent, the valuer needs also to assess the actual characteristics of the land in a similar manner to vacant land as mentioned in Section 2.3.1 even though these may not be a direct market consideration for the hypothetical purchaser.

4.3.7 Type of Cropping

The type of farm enterprise can in some ways have a similar implication for the investment market as land classification. It has already been established that different categories of investors have different preferences with, for example, the financial institutions having favoured larger blocks of commercial farms on higher grade land. When such investors were particularly active, there would be increased demand for big arable let farms with a resultant upward effect on their values. Conversely, if a large and well let dairy farm were being offered for sale, it might have attracted less interest from the institutional sector mainly because it would not be recognized so easily as being a suitable investment.

Such factors may however have the opposite effect on some investors. The traditional institutions have been known to buy land that contrasts with the bulk of their existing holdings in order to create an element of balance within their portfolio. The Duchy of Cornwall whose main holdings are in the south-west with a preponderance of dairy farms and smaller units have, for example, been contenders for large arable investments in East Anglia just so as to maintain an interest in another part of the agricultural economy. As valuation depends to some extent upon interpreting the market at the time, it is important to know when exceptions to the normal pattern might occur.

4.3.8 Fixed Equipment

The significance to freehold value of farm buildings and other items of capital equipment was considered for vacant land in Section 2.2.3. These same considerations apply also to let farms in that the existence of a facility such as a grain store will enhance the tenant's income in the same way as in the previous case of an in-hand farmer. One would expect that the opportunity for gaining this improved income should then be reflected in a better level of rent and so also increase the investment value. This may not always be the case, however, as it is possible that all or part of that capital equipment has been provided by the tenant. In that event it will hardly be feasible for the landlord to include it in his rental calculation and the valuer when making his appraisal will therefore need to check whether such arrangements exist on any buildings or fixed equipment.

In most cases the buildings will be part of the freehold property and belong to the farm. It can happen that a tenant has had to add to the original farmyard at a time when further covered space was required by the farm business but when the landlord was unable or unwilling to finance the building work. This would normally be for such items as a lean-to extension to a covered yard or new milking equipment but can also be for a complete building such as a grain store or a specialist range of structures like poultry houses. Traditionally landlords were perhaps in a position to accept that capital invested in their estates would show only a modest annual yield of around four per cent or less. If then they were in a position to fund a new building they would do so against that kind of return which would be receivable in the form of an increase in rent. In a more modern context, however, landowners would only be able to justify the expenditure if they could cover the annual cost of finance, which is likely to have been at least twice and probably over three times the agricultural investment rate. If this were to be covered by an additional rental charge, the tenant would be faced with a massive and possibly unsustainable increase over the prevailing level. Even if this high rent were to be accepted, the landlord could still be at risk of facing a notice for a downward negotiation at the time of the next statutory review. That situation might be in fact be avoidable if the interest charge were to be agreed under a separate arrangement to the original agricultural lease. None the less, where the tenants may have some of their own capital to invest or are able to benefit from lower construction costs by doing some of the work themselves or with farm staff, it may be that the only answer will be for both parties to agree that the tenants undertake the project themselves

Having identified that buildings have been added to the farm subsequent to the date of the original agreement, the valuer then needs to establish a number of points. If the landlord funded the building against an increase in rent, one must assess whether that could still be reduced, as this would affect the reversionary calculation which is set out in Chapter 5. This will be a matter of comparing the prevailing total rent against the levels that would be judged to be appropriate in the current local market. It is possible that the landlords avoided this possibility, or at any rate reduced the risk, by making the interest charge on the new building the subject of a separate agreement independent of the main lease. All this should be evident on reading through the formal agreements made between the two parties. If the tenants constructed the building, they will have been advised to have written landlord's consent and also agreed a basis for compensation should they vacate the farm during a specified period, which will be equivalent to an accepted financial lifespan of the premises. On the face of it, this seems unlikely to affect the value of the property. First, statutory tenants have such long-term security that it would be unusual for the lease to be terminated except by some misadventure. Second, if possession were obtained, the value would rise to the full vacant level and this would eclipse any financial commitments that the landowners might have to their former tenants or their beneficiaries.

On this last point one should be careful not always to presume that the owner would be realizing funds by selling the farm as it became vacant. It may be that the land is situated within an estate or portfolio so that it could not be sold off without detriment to the remaining holding. In that case the owner would need to find the amount of the agreed compensation. It is possible that in such a situation, where a major improvement has been carried out relatively recently so that a substantial amount of compensation would apply, that a hypothetical purchaser would have reservations about taking on this potential commitment and that the offered price and therefore the value might be adjusted accordingly. Again, one might argue that the chance of the tenants quitting the farm is sufficiently remote for such adjustments to be unnecessary. However, where the tenants have committed themselves to a substantial investment then the business could initially be under a greater pressure and so at greater risk, therefore making this issue potentially more significant.

4.3.9 Redundant Buildings

There is a further legal consideration that the valuer must take into account when inspecting the fixed equipment on a tenanted farm. It is possible that some of the older buildings will have been scheduled formally as being redundant. This will have been agreed with the particular intent of releasing either party from any further obligation to maintain those buildings as would otherwise be required under the overall terms of the lease. It means of course also that the landlords will from then on have to exclude those buildings from their rental assessment, which may have an impact directly upon the freehold valuation. Under the 'model' clauses, it would also mean that they would no longer have to allow for the cost of future maintenance which would be likely to detract from the value of the whole as money would have to be set aside for such anticipated expenditure. It does not necessarily mean however that the tenants will no longer continue to use them and even to repair them to whatever standard they think fit. Where there are old buildings still in use on a farm, the valuer will need to check the tenancy agreement to see whether they have been scheduled as redundant. Where this is the case, allowances must be made for the appropriate level of rent and the assumed cost of maintaining the buildings. Where they are not redundant, the repairing costs will have to be reckoned either as an outgoing for the landlord or as a factor reducing the rent payable by the tenant, according to who has that responsibility. If, in this latter case, one were then to assume that an investor might seek to avoid these issues by offering to agree to making the buildings redundant, the farm could then be deemed to be short of the necessary facilities. The value would then be affected accordingly, as mentioned at the beginning of Section 4.3.8, and one would be looking at the implications of providing a modern alternative.

4.3.10 General Condition

As in the case of vacant properties, the value of a tenanted farm can also be influenced by the general condition of the land and buildings although often to a differing degree. With in-hand farms this is determined partly by the assumed cost of putting the premises into good order and partly by overall market attitude (Section 2.3.7). Where the property is let on traditional terms under which the landlord has direct responsibility for some of the repairs, the same attitude could well prevail, and the value discounted by the anticipated cost of carrying out the necessary works. Under modern full-repairing terms, one may be tempted to assume that the prospective landlord can ignore such costs as he can expect them to be borne entirely by the tenant. This may however be unrealistic on two counts.

First, even if the tenant is responsible for maintaining the buildings, the fact that they are in need of significant expenditure may affect the level of rent that would be otherwise achievable from the farm. Then, the fact that the land and buildings have been allowed to fall into disrepair can suggest that the tenant is not sufficiently diligent or competent and that past management has been a little lax. The calibre of the sitting tenant is of course a major part of the investment value (Section 4.3.5) and the condition of his farm will naturally have a bearing on this. One should be aware, however, of circumstances in which old buildings may be in poor order but where this is not a reflection on the ability of either the tenant or of his landlord's management. The cost of maintaining some traditional buildings can be more than would be justified by either the profitability of the farm or by the investment return to the owner so that there is little alternative than to allow them to fall into bad repair.

The implications for the value of the property as a whole will then focus more on the constraints on rent while the farm is without any modern alternative accommodation and on potential liabilities for the owner if the building were to be listed as being of architectural and historic importance.

The other distinction between how the general condition of the farm affects vacant or tenanted values arises from the attitude of the hypothetical purchasers. Where in-hand land is being bought by farmers for their own occupation, they will tend to take a commercial approach and assume that they will be able to put the property into the standard of repair that they think appropriate and that this will only be a matter of allocating the necessary time and finance. Many investors, on the other hand, will set greater store upon the property looking well and will be more put off by one that is in bad order so that the price may be discounted to a greater degree than in the case of a vacant farm. The effect that the condition of a tenanted farm may have on its value will therefore depend to some extent on the market within which it lies, as outlined in Section 4.2 above.

4.3.11 Rent

The level of rent being paid is a significant element in the valuation of a tenanted property, although there is little distinction in the way that this might influence the various kinds of investors. This is due to the fact that rents can be reviewed at quite frequent intervals and that it is therefore unusual for any to lag much behind the general market trend and so offer a particular reversionary uplift that might appeal to a certain kind of buyer. The subject of rents is dealt with in more detail in Chapter 6.

4.4 Development and Diversification

The general implications for developing farmland or adapting it to other uses are dealt with in Chapter 9. In this present context, it would be appropriate to consider how such matters may affect tenanted properties.

4.4.1 Development

As mentioned in Section 4.3.5 above, whilst tenants have a statutory right to a continued exclusive occupation of their farms, landlords do have the opportunity to gain possession should they obtain consent to use the land for other purposes. This right is coupled with a provision to pay compensation to the tenants and the level of compensation is of direct consequence to the valuation. In such cases where development was anticipated at the commencement of the tenancy, it is possible that the lease may include special provisions for early entry by the landlords.

Under the statutory regulations, landlords are required to pay to tenants who vacate all or part of their holdings an amount equivalent to five or six years' rent apportioned to that area of the let holding actually taken for development. The actual sum is defined in two parts; under Section 60 of the *Agricultural Holdings Act* 1986 an amount of between one and two years' rent is specified and an entitlement to a further four years arises under Section 12 of the *Agriculture (Miscellaneous Provisions) Act* 1968. To this must then be added any compensation due to the tenant at the termination of the lease, in respect of, for example, capital invested in improvements in the past, as well as an amount for tenant right relating to certain shorter-term expenditure incurred during recent farming operations and which still has some unexpired value. The valuation of tenanted properties with development rights will therefore differ from those with vacant possession in that such sums as are due to the tenant will need to be deducted from the capital value assessed by the means set out in Chapter 6.

There can also be situations where the landowner may not trust to the statutory provisions or where these would involve a longer process than would be acceptable to the developer. In those cases a landlord will need to rely on gaining possession by negotiation which is likely to involve a larger amount of compensation. Even with a legitimate requirement to take the land in hand, the landlord will need to give a minimum of 12 months' notice commencing on the next term date of the lease. It would therefore take between a minimum of a year and up to two years to gain possession by the official formal means.

4.4.2 Diversification

It has become increasingly common for part of a farm to be given over to some alternative use other than agriculture. When this is implemented by the landlords, the valuation implications will be similar to any other development as described in Chapter 6. If however, as can often arise, the diversification is carried out by the tenants, then other factors must also be taken into account.

Normally tenants will have obtained formal consent from their landlords to carry out the relevant works such as converting some traditional buildings. In obtaining this consent, the tenants will probably have secured an agreement to a formula for compensation from their landlords should they vacate the farm within the specified economic lifetime of the improvements. In most cases the valuer may assume that the tenants are intending to remain on the farm for the long term and that such compensation is not be likely to be demanded and need not therefore be incorporated in the valuation. The only occasion when the tenants or their representatives might have a claim for compensation would be if the lease were unexpectedly terminated through some unforeseen event such as the death of a tenant or perhaps the collapse of the diversified business. In such an occurrence, the landlords would be gaining the vacant possession premium over the whole and might not therefore need to worry about paying any outstanding compensation. If, however, the landlords were then unable or unwilling to realize the vacant value through a sale, they would still need to budget for the sums due to the tenants or their successors. Although it is unusual for landlords to face such a situation, it may need to be allowed for in a valuation particularly if the scheme has involved a substantial amount of capital or carries an element of commercial risk. The amount by which the valuation might be discounted for this will rest more upon an interpretation of how in a hypothetical sale the purchasers would wish to compensate for the potential risk rather than on any specific calculation.

Check-list – Land let on Statutory Tenancy:

- Tenanted land as an investment and assessed in terms of financial yield.
- Yield determined by market factors, with demand arising from investment funds or individuals and from sitting tenants.
- Different investment requirements and need to identify to which type of buyer property is likely to be appropriate.
- Implications of terms of lease, on landlord as investor and on rental return.

- Security of tenure and possibility of vacant possession premium, depending on circumstances of tenant.
- Land classification as a measure of investment quality.
- Quality of tenant, and of land and fixed equipment, and arrangements for redundant buildings.
- Development and diversification as a partnership between landlord and tenant.

CHAPTER 5

Investment Calculations

The concept that the worth of a tenanted property can be expressed in terms of its yield was introduced briefly at the beginning of Chapter 4 and now needs to be looked at in a little more detail.

In its simplest form, when a farm is let under the provisions of the *Agricultural Holdings Acts* the tenant is entitled to remain on the holding for his lifetime and, in certain instances, for the lifetime even of his successors, provided that he fulfils his obligations under the terms of the lease. In the most straightforward cases, one may assume that the tenant or his family are likely to remain on the holding for many years to come and that they will be competent enough to meet their legal obligations and in particular to continue paying the required rent during that time. This means that the landlord will be assured of receiving that rent for an unspecified, but presumably lengthy, period which for the purposes of investment calculations is considered to be equivalent to perpetuity. When there are uncertainties or potential variations in this arrangement, the receipt of income is assessed for a specified limited term and an allowance is made for the changed situation that follows, known as the **reversion.**

Referring, however, for this first example to the case of an uninterrupted on-going tenancy producing a full market rent, the landlord or investor would need to establish what in capital terms it would be worth to him today to be able to receive that rental income effectively for evermore. Any capital investment will need to show a rate of return on the sum expended and that rate is determined by a combination of factors relating to the perceived quality of the property, the security of income and of capital value and the potential for growth. This can be expressed as follows:

$$\text{income} = \text{capital} \times \text{rate of return}$$

As the rate of return is expressed as a percentage figure, if this were 5 per cent, the example becomes:

$$\text{income} = \text{capital} \times \frac{5}{100}$$

or alternatively as:

$$\text{capital} = \text{income} \times \frac{100}{5}$$

This inverse of the rate per cent is known as the Years' Purchase, or YP, and would in this simple example be 100 ÷ 5 which equals 20. When there are grounds for expecting interruptions to the long-term income flow, the calculation of YP becomes more complicated, but in this most straightforward form capital value can be deduced by multiplying the income by the inverse of the rate per cent that is deemed applicable to that particular investment.

This principle may be applied to a straightforward example. Consider a farm that is let on a statutory agreement at a newly reviewed rent of £15,000 per annum to an able tenant aged about 40. The property is of a type and quality for which one would in the present market assume an initial net yield of five per cent. As no change is foreseen in the prevailing arrangement, the YP is taken to be in perpetuity and is therefore the inverse of four per cent or 25. This multiplier has to be applied to the income received by the landlord and can be reckoned either as the gross rent paid or the net amount left over after the landlord has met the annual cost of ownership such as management and possibly repairs and insurance. Which position one takes will depend on what interpretation was made about the assumed market yield. Strictly speaking, this should represent a return from the net income after all costs, although there has also been a tendency to work with gross rental figures particularly at a time when there may be scant market evidence and as more leases are on modern terms.

Taking the gross rent, the calculation is simple:

i) *Basic capitalization of a rack rent.*

Rent	£15,000	
YP @ 4% in perpetuity	25	
Capital value		£375,000

If however, the assumed yield of four per cent was one that investors would apply to the net income that they receive from the capital sum invested in their holdings, one would need still to make adjustments for the costs of purchase and management before applying the multiplier to capitalize the rent.

The capital value is after all the figure at which a hypothetical purchaser would buy the property. When assessing the amount that he would spend and the return that he would achieve, one must allow for any extra costs that are incurred over and above the purchase bid. These will be solicitors' fees for conveyancing and also agents' commission where a purchaser may have used such professional services in the acquisition, as well as tax in the form of stamp duty. For valuation purposes, these will be approximated as a percentage rate of the price, currently around 4.75 per cent, assuming stamp duty at three per cent, conveyancing costs at around 0.75 per cent and agents' fees at one per cent.

When the property is let on modern terms with the tenant responsible for all repairs and insurance, the landlord's only regular outgoings are management fees that will be on a relatively modest scale as the agent is not involved in detailed matters of maintenance. One may assume in most cases an average figure of five per cent of the gross rent. Incorporating both purchase and management costs, the calculation then becomes:

ii) *Capitalization of rack rent after allowing for management and purchase costs.*

Rent	£15,000		
less: management fees @ 5%	£ 750		
net annual income	£14,250		
YP @ 4% in perpetuity	25		
Gross value		£356,250	
net of purchase costs @ 4.75%		÷ 1.0475	
Capital Value			£340,095

This calculation demonstrates the full range of adjustments that should be made if it is clear that the investors are taking such a precise view on yield. In practice, however, the market may also work with gross yields, particularly when dealing mostly with land let on modern terms or at times of only limited evidence. The allowance for purchase costs are of more significance when assessing a purchase bid, as the investor will need to budget for this expenditure over and above his top price. In yield terms it may however often be assumed to be included in the investment figures, especially as the acquisition fee and tax rates are broadly similar for all types of property within the separate market sectors. On the other hand, as the different types of leases involve different management liabilities, the allowance for annual outgoings is more likely to be incorporated in an investment appraisal provided that there is sufficient market evidence upon which to make an accurate assessment of yield levels.

Therefore, where the farm is of a traditional nature and let on the old 'model' clauses with the landlord responsible for main repairs and insurance, a further deduction has to be made firstly for the estimated annual cost of maintenance and repairs and secondly for the appropriate rate of management fees. The former will depend on the nature and extent of the buildings and their general state of repair. The latter will reflect the likely time spent on overseeing these items. As a very rough guide, expenditure on maintenance and insurance may be taken as about 15 per cent of the gross rent and management fees will be around 7.5 per cent and 10 per cent. This would be represented as follows:

iii) *Capitalization of rack rent on 'model' clauses.*

Rent	£15,000	
less: repairs & insurance @ 15%	£ 2,250	
less: management fees @ 10%	£ 1,500	
net annual income	£11,250	
YP @ 4% in perpetuity	25	
Capital Value		£281,250

In practice, the yields may vary according to the type of lease and the level of outgoings. This will depend upon the attitude of investors at the time but it could either be that there are circumstances when they would accept a lower net annual yield on the more traditional holdings or, on the other hand, that they will need to see a slightly higher return to compensate for the inconvenience and uncertainty of having the responsibility for repairs.

There is of course another deduction that landowners may have to make from the gross rent, namely income or corporation tax. This is however not normally done in valuation calculations and, indeed, some forms of investors enjoy tax exemptions. The 'net annual income' is therefore still gross of any tax that may be payable. It is possible that the owner may be registered for VAT but this is also ignored in these investment calculations. Landowners do in fact have the right to exemption from VAT and may elect to do so, primarily to avoid the resulting limitation that could occur if the property were to be sold in the future to investors who are not registered for VAT.

Having established the appropriate current net income figure for capitalization, one then needs to consider whether this is an up-to-date market level or whether there would be scope for it to be reviewed on the next term date. Under the *Agricultural Holdings Acts* either party may serve notice to review the rent at intervals of not less than three years. In these statutory agreements, there is opportunity for the rent to be reduced as well as increased and so the valuer will need to consider both possibilities, in contrast to many commercial property leases where the reviews can be on an upward only basis.

One must therefore assess the level of rent that would be applied to the holding as at the date of the valuation, using the principles outlined in Chapter 6. This figure is assumed then to be that which one would secure at the date of the next review and this is built into the capitalization by the means outlined below.

Where the property involves a more diverse tenure than the previous illustration of an ongoing long-term tenancy the YP cannot be derived by simple division. Although the principle remains the same, the calculation would involve a more complex formula and it is more practical to refer to special tables. The most common source of reference of these is in a book known as *Parry's Valuation and Investment Tables* in which are set out tables giving the figures for YPs and other multipliers for specific periods of time. This then enables one to capitalize rents that are due for review or to work out the present worth of a reversionary opportunity. The kind of functions that are more commonly encountered in the agricultural investment market can be demonstrated by considering the following examples.

Dealing first with the scenario where the rent is likely to be reviewed at the next term date, the present rent is capitalized as if in perpetuity and as in the previous examples. If the current open market rent, however, is assessed as being, say, £17,500 and the next review date is two years from the time of the valuation, then the anticipated increase of £2,500 will be capitalized as an additional item whilst being deferred by the period of two years:

iv) *Allowing for rent review.*

Current rent	£15,000	
less: management fees @5%	£ 750	
net current annual income	£14,250	
YP @ 4% in perpetuity	25	
		£356,250

Rent in 2 years	£ 17,500	
less: management fees @ 5%	£ 875	
net annual income in 2 years	£ 16,250	
less initial net income	£ 14,250	
net increase in income	£ 2,375	
YP @ 4% deferred 2 years	23.11	
		£ 54,886
Capital Value		£411,136

This calculation can also be done by capitalizing the present rent for the remainder of the present term, namely two years, and then applying the appropriate deferred YP to the anticipated reviewed rental figure, as follows:

v) *Alternative capitalization of rent review.*

Current rent	£ 15,000	
less: management fees @ 5%	£ 750	
net annual income	£ 14,250	
YP @ 4% for 2 years	1.89	
		£ 26,932
Rent in 2 years	£ 17,500	
less: management fees @ 5%	£ 875	
net annual income in 2 years	£ 16,625	
YP @ 4% deferred 2 years	23.11	
		£ 384,204
		£ 411,136

Whilst the two methods in examples (iv) and (v) produce the same result, there is a different emphasis. In the latter case the greater part of the valuation figure stems from the reviewed rent and, if there had been reason to use a different YP for the assumed reviewed rent, the two calculations would have produced different results. The question of whether to use a lower YP for such future positions is considered under the example (vi) below. In situations where the rent is likely to be decreased, it would be more logical to use the methodology illustrated in example (v).

The same general principle is applied to situations where the lease is likely to be terminated. Whilst this is of course rare for statutory tenancies, it can arise with longer-term Farm Business Tenancies, as mentioned in Chapter 7, or when land is given up fc where there is an older tenant who has no obvious e introduces the concept of reversion. Taking a simple ex ern fixed-term tenancy that has another eight years to ı will be receiving rent for that time after which the far. taken in hand and producing a direct income or it cou vacant possession. For valuations, it is normal to look to the latter scenario as this would produce the more definite capital position.

The calculation begins as in the immediate preceding example, but the deferment is then to the present estimated capital value of the property rather than to a capitalized reviewed rent :

vi) *Reversionary interest.*

Current rent	£ 15,000	
less: management fees @5%	£ 750	
net current annual income	£ 14,250	
YP @ 4% for 8 years	6.73	
		£ 95,902
Present vacant possession value	£550,000	
PV of £1 in 8 years @ 4%	0.73	
		£ 401,500
Capital value		£ 497,402

If there were still within the remaining eight-year term an opportunity for reviewing the rent, then this could be incorporated into the above calculation as in the examples (iv) and (v) above.

There is however also a question of which yield to apply to these calculations. So far the above examples have all used the one rate of four per cent that was deemed at the beginning to be the appropriate figure for a tenanted property of this kind in the prevailing market. One of the factors that determine the accepted rate of return is that of security. Where the income yield on an investment is considered to be secure, not only in terms of receiving the present rent but also as to the likelihood for growth, this will command a lower rate than another proposition where this may not seem quite so reliable. Strictly speaking, an anticipated future rent cannot be so secure as one that is currently being paid. It is after all possible that unforseen events may occur between the present time and the date of the next review which could influence the level of rent that could then be achieved. This potential uncertainty might have to be recognized by applying a slightly higher rate of return to that part of the calculation, as follows:

Dual rates for deferments.

The amount by which the rate should be increased in such a case has to be a matter of judgement. In the example of a rent review in two years' time, there is a relatively short period intervening, during which it is perhaps unlikely that anything unexpected would still occur to affect the estimated level of rent review. It may therefore be unnecesssary to try to make an allowance for what would be considered a fairly limited risk. There may be other deferments however where there is a real element of uncertainty such as where planning consent is being sought for a future development and where an investor would need some protection or comfort against the possibility that it may be delayed or not granted at all. One should recognize too that where there is a longer or less precise period to the reversion, as in the case of anticipating the death or retirement of a sitting tenant, there is perhaps a greater chance that something might change or go wrong during the intervening years, and that ought then to be reflected in a slightly higher yield.

In practice, the adjustment to the yield will be more a matter of showing that one has recognized the possibility of some uncertainty in the future level of rent or capital reversion rather than a precise quantification of the degree of risk. In the example where the current yield is assumed to be around four per cent, the adjustment could range from nil in the case of a normal rent review to between about 0.25 per cent to one per cent for a situation covering a longer period or less certain outcome. An example of this is given in (viii) below.

The choice of yield applied to deferments involves another consideration too. The deferment is a 'Present Value' of either a capital sum or of an income, both of which will only arise at some future date. In the above example (vi) it was seen that being able to gain vacant possession in eight years' time was worth considerably less than the amount that would have been paid for such a property at the time of the valuation. This discount is not so much a measure of risk but an assessment of the opportunity cost to the landlord in not having access to that capital sum until the end of the lease. It could be argued then that if the owner were able to sell with vacant possession today he could either invest the proceeds in a higher-yielding sector such as the money markets or use them to reduce borrowings and thereby also the associated cost of interest. In both cases the alternative opportunity relates to money rates rather than property yields and these are generally higher than those achieved in the farmland market. This suggests that one should use an equivalent of a money rate when capitalizing a deferred interest rather than one reflecting purely the property features.

Ultimately, it must reflect as far as possible the market attitude and whether purchasers for a particular type of proposition will be assessing it on the basis of a financial opportunity cost or purely as a property investment. It may be then that in some instances, such as land being bought for development or on a fixed-term tenancy, it will be appropriate to use the equivalent money rate. In most conventional cases, however, the relevant property rates are to be applied throughout the period of ownership, and any comparison with money rates would ignore the fundamental factors of the land market, such as gaining capital as well as income growth, which justified the use of the lower investment rates in the first place. A convenient measure of what has been described here as the 'money rate' can be that offered at the time on Government Gilts.

This equivalent of a money rate can be used in one alternative approach to the question of future reversions and rent reviews. The element that is considered to be at risk is capitalized separately at this higher rate whilst the part that is accepted as being within market norms is calculated at the appropriate property rates, as in the preceding examples. This technique known as 'top-slicing' is however more relevant to certain commercial property situations and is rarely used for farmland.

In this context it is worth noting that the concept of risk is different when applied to agricultural as opposed to commercial property. In a farming situation, if rents become unsustainable at some future date the tenants may have the opportunity to negotiate a reduction at the next review with a resulting fall in return to the landlords. If however the tenants were unable to pay the rent, they would be forced to leave the land so that it then reverts to full vacant possession value with a resulting capital gain, or reversion, for the landlords. For many commercial leases, on the other hand, landlords may be protected from the possibility of a downward review, but if the tenants were to default then the property would become vacant and probably suffer a fall in value until such time as it could be relet on a full rent.

Another calculation in deferment that can arise with agricultural property is that of life interests, either when having to estimate the likely termination date of a statutory tenancy or when a farm or estate is owned or occupied specifically for the duration of one life or even a combination of lives. The principle is the same as that used for a fixed-term lease as in (vi) above, except that it is not clear as to the period of years for which one should take the Present Value of £1 (see Glossary page 197). The answer, for valuation purposes at least, can however be found in Parry. A section towards the end of that book is based upon the official statistical figures for life expectancy which can be applied as follows.

viii) *Life interests.*

In this example the farm is let on a life tenancy to a man aged 62. The Life Tables show the average mean expectation of life as 15 years. The calculation therefore becomes:

Current rent	£ 15,000	
less: management fees @ 5%	£ 750	
net current annual income	£ 14,250	
YP @ 4% for 15 years	11.12	
		£ 158,460
Present vacant possession value	£550,000	
PV of £1 in 15 years @ 4.5%	0.517	
		£ 284,350
Capital value		£ 442,810

The higher yield rate of 4.5 per cent for the latter part of the calculation is effectively an allowance for the degree of uncertainty that arises, in this case from two causes. First, the assumed time span of 15 years is only a theoretical estimate and far less reliable than, for example, the termination of a formal lease. The landlord may therefore have to accept a far longer period of tenanted occupation, if the lessee were to live longer than the average. On the other hand, the chances that he may die earlier are statistically more remote and as an investment principle it is unlikely that that could give one grounds for accepting a lower reversionary yield. The second issue is that 15 years is a relatively long time during which there is a correspondingly greater possibility of unforeseen changes occuring than in the earlier example (vi) of just 8 years remaining. Such changes, whether in the market or to the property itself, could of course be as easily in a landowner's favour as to his disadvantage, but again as an investment principle this represents uncertainty and as such will tend to be reflected in the application of a higher yield.

ix) *Succession.*

In contrast to the above, situations do arise where the property is likely to remain in tenanted occupation for longer than normal, in that the current tenant's heirs may be entitled to claim successionary rights to the lease (Section 4.3.5). In the original example (i) above, the letting to a tenant aged 40 was assumed for the purposes of calculation to be in perpetuity. When however a succession is anticipated so that the lease may be prolonged beyond a single lifetime, one can hardly extend the investment term beyond perpetuity. None the less, the fact that there is a potentially longer term of occupation is likely to affect its price as investors reckon that with a longer period the vacant possession premium could be achieved. This can be recognized in the valuation by using a slightly higher yield to reflect the lesser attraction of the investment whilst still using the same calculations assuming perpetuity.

x) *Market changes and inflation.*

The figures used in these forward investment calculations are the equivalent of market levels prevailing at the time of the valuation even though it is more than likely that prices and rents will have changed by the time of the review or reversion. In the above example (viii) one may well expect that in 15 years' time the vacant possession value will be very different to the present one. After all, looking at the graph of property values given in Figure 4.4, one can see that there had been considerable fluctuations in average investment values over the last recorded 15-year period. It is not, however, the function of the valuer to predict future changes whether in market movements or in inflation. An investor may well take a view on such matters and refer perhaps to forecasting models, and these considerations could then influence the level of current yield that he would expect from that particular type of property. The values to which those yields are then applied will still have to be an assessment of current levels. This does not mean that the valuer may overlook foreseeable changes in the property itself that would affect future value. For example, if some farmland has alternative potential that could be exploited at the end of a tenancy, then one would take the current market value appropriate to that enhanced use and incorporate it in the figure against which the deferred yield is applied. It is by now clear that the actual yield used will reflect the degree of uncertainty assumed in achieving that additional value in the future but the capital figure will none the less be that prevailing at the time of the valuation for the property as if it were able to benefit from the alternative use.

The wider principles regarding the question of current figures being used in a constantly changing market are defined in the Red Book as mentioned in Section 1.3.

There are some further items in Parry's tables that are rarely used for agricultural property but which should be mentioned briefly.

xi) *Sinking Funds.*

With certain forms of property interests, such as leaseholds, the owner's rights diminish over time until they revert to another party at the end of the term. In a case of this kind, the rental income may have to fulfil a double function; first, providing an annual return on the capital invested and, second, building up a sinking fund to replace that initial capital at the end of the term. The sinking fund is likely to be calculated at a different rate to the property yield, being dependent on the financial markets, and the tables therefore give a combination of two selected yields, or a Dual Rate YP, over a range of terms up to 100 years. Since the sinking fund contributions will normally be made out of taxable rental income, the tables do in this instance allow also for tax being deducted at certain appropriate rates. As leaseholds are rare in rural property, these tables are not often needed in a conventional farm or estate context, although they might be applied for investments made in diversification schemes that are expected to have only a short commercial life.

xii) *Discounted Cash Flows and Internal Rates of Return.*

Agricultural investment valuations tend to differ from those in most other property markets in that the expected rent changes or capital reversions are comparatively simple. Investors are therefore inclined to define their requirements in terms of initial yields achievable at the time of purchase or possibly by reference to the yield anticipated at the first review which will normally be within a relatively short time. The calculations described in the preceding sections therefore fulfil most market situations and there is little recourse to the kind of valuation methods used commonly for other types of investment property. There can be occasions, however, even in the rural sector, when a property is likely to experience a more diverse pattern of income or capital changes, and when its current value is best assessed by means of a more sophisticated approach incorporating Discounted Cash Flows (DCFs) and Internal Rates of Return (IRRs).

In a commercial property context, these will tend to be calculated from a specialized computer programme, although IRRs are in fact also given in Parry's. The principles involved in these calculations are basically an extension of the concept of 'Present Value' as used in each of the preceding examples. Discounted Cash Flow is a measure of the current value of an anticipated future stream of income (and expenditure) discounted back to the present time at the appropriate investment rate. In the context of what has been described previously in this Chapter, it is the sum of a series of calculations of Present Values of £1 per annum, made feasible by computer programming. The Internal Rate of Return is a means of establishing what yield this net income flow represents over the term of the investment and against a specified purchase price. This enables one to make comparisons between investments that have different patterns of income and expenditure but are otherwise sufficiently similar to be assessed against each other. It can also establish a purchase price, or current value, once it is known what yield would be warranted by the proposition within prevailing market conditions or would be sought by a particular investor. Reversions in the agricultural property market are generally clear enough to be accounted for by the conventional means described earlier in this chapter, although there may be sectors such as land with development potential where investors might judge the market level by reference to the IRR.

Although still rarely required for most agricultural properties, these more commercial methods of valuations may have an increasing relevance in the rural field, whether for farms involved with diversification schemes or where land is linked to increasingly complex development agreements. IRRs are also already used in the appraisal of commercial forestry, as outlined in Section 10.5.3.

Further details on these methods are best obtained from books dealing with commercial property valuations, as noted below in the appendix on further reading.

For a checklist on Investment Calculations, please turn to next page.

Check-list – Investment Calculations:

- Basic principle of capitalizing income and selecting rate per cent or YP.
- Use of Parry's Tables.
- Allowances for perpetuity, limited interests and reversions.
- Deferred interests and dual rates.
- Inflation, forecasting and IRRs.

CHAPTER 6

Tenancies

6.1 Rental Valuations

The rents that are paid to a landlord create the income on which the investment return is assessed. It is important therefore to be able to establish whether the current rent is at a proper market level or whether any allowance should make for a review as shown in the worked examples (iv) and (v) in Chapter 5. It may be sufficient in some of the more straightforward cases to make just a general comparison with similar farms in the locality, but there will also be instances when there are too many special factors to consider and too few comparables so that more detailed appraisals will be needed.

The general approach may apply in a case such as a commercial arable area like the Fens, where the value can be determined largely by the nature of the land itself and where the houses and buildings provide little variation. If it is known that rents for that kind of land tend to be at, say, £200 per hectare then this can be an adequate benchmark for other similar properties given that they do not have too many distinctive characteristics.

More often, however, there will be variable features that will need to be taken into account so that a more detailed appraisal will be necessary. This will involve two functions: first, an assessment of the quality of the property itself in very much the same way as for a valuation of vacant possession land, as described in Chapter 2, and second, an estimate of what the farm may be expected to produce. How these appraisals can then be used to establish the proper rent will depend on the type of lease and upon a number of assumptions and is dealt with later in Sections 6.1.2 and 6.1.3.

The financial performance of a farm can be determined by means of preparing a budget for the current business. One would expect this to involve a highly technical exercise requiring not only direct farming expertise but also a close familiarity with the land and its buildings. Essentially this is of course the case, but there are however ways in which a suitable form of budgetary analysis can be produced even by those with only a more general knowledge of such matters. The farming figures can be obtained from an accepted management reference book giving prices and costings for almost all the types of enterprise that one might encounter. The one that is most frequently referred to in this context is the *Farm Management Pocketbook* by Professor John Nix of Imperial College at Wye and which is now published annually by Andersons of Melton Mowbray. The equivalent data are also produced by the Scottish Agricultural Colleges in their *Farm Management Handbook* in which the figures are prepared specifically to reflect performance levels in the northern half of Britain.

The use of such information may be helpful in establishing rental values but one must recognize that it cannot provide all the answers to farm profitability, especially for anyone without proper agricultural training. Furthermore, it would not come within the remit of a book on property valuation to explain fully the workings of farm management tables. None the less, it would be useful to have a brief explanation of this method, as follows.

6.1.1 Budgets and Gross Margin Analysis

Conventionally, farm businesses are assessed on the basis of Gross Margin analysis. This involves establishing the gross output from each separate enterprise on the farm and then deducting the associated variable costs to arrive at the gross margin. The gross margins are then added together and from that total is then deducted the fixed costs for the entire business to give the net farm income, generally prior to any deductions for interest or for personal drawings. Gross output is calculated by the price per unit multiplied by the number of units, whether hectares or head of livestock, and then by the price per unit. Variable costs cover the cost of inputs for that enterprise, such as seeds and fertilizer or bought-in feed. Fixed costs are those that are carried by the farm as a whole such as wages, machinery repairs and depreciation, and fuel and services. In essence this is an exercise in deriving a net income by reference to accepted data for calculating gross receipts and deducting all appropriate production costs.

To prepare a comprehensive farm budget requires a comprehensive knowledge of agricultural systems and reference to a relevant business management text book, such as *Financial Management for Farmers and Rural Managers* by Martyn Warren. However, for the present purpose of gaining an understanding of rental valuations it may be helpful to illlustrate the method by the simple example given in Figure 6.1.

All the figures used in that example are taken from the Nix, *Farm Management Pocketbook*, which is a recognized standard for the industry. For each enterprise, however, this book gives four alternative figures representing: average, high, medium, or low. The reader will then need to know which of these to select for the farm in question and this will depend on the quality of the farm and the calibre of the stock or of the management. The quality of the farm is judged much as for a vacant property valuation, but the potential of the stock and the policy aspirations of the management involve a wider range of skills. Past performance may not provide an accurate indicator of which category one should use for budgeting purposes. Poor yields in previous years may be due to bad management or inadequate in-puts rather than to the practical limitations of the farm itself. Conversely, high yields might have been achieved in the past by a policy of costly in-puts such as fertilizer or labour, so that the financial results will turn out to be more modest than the actual physical output would suggest. Therefore, when preparing a budget one has to look separately at every stage of the farming business to be sure of arriving at the correct net final position.

The example given in Figure 6.1 is in its simplest form which may suffice for certain types of rental valuation but there will also be occasions, particularly in rent negotiations, when a more detailed approach will be needed. Each element of the Gross Margin Analysis can be worked out for itself rather than taken as the overall per hectare figure. Some of these calculations are even given in the *Farm Management Pocketbook*, such as working out gross output through multiplying crop yield by price or showing a breakdown of variable costs.

Even in a straightforward analysis as that in Figure 6.1, there will be technical considerations to make. In the case of cereals, it must be decided whether the crops are being produced for the lower-priced feed market or for the premium levels of milling wheat or malting barley. The assessment of a livestock enterprise follows the same principles as in the arable example but there will again be details to determine with regard to feeding and breeding regimes.

The figures that are given in the Pocketbook for prices and costs are prepared one year in advance and are generally accepted for budgeting purposes for that period. Many of these remain relatively unchanged during that time, but there will also be situations where it would be more appropriate to refer directly to current market evidence and to incorporate such figures with those taken from the book. These prices can be obtained directly from sources such as corn merchants, livestock auctioneers or milk processors, but many are also well recorded each week in the main farming papers, as itemized in Section 3.1.

This raises the question of the date for which a rental valuation should apply, in a similar way to that for property valuations as mentioned in Section 1.3. When prepared in conjunction with a rent review, this will involve considerations of the period over which that rent will

Figure 6.1 Gross Margin Analysis

Arable farm of 400 hectares in England being cropped with a rotation of winter wheat, winter barley and spring beans.

Crop	*Hectares*	*Gross Margin per hectare*	*Gross Margin per enterprise*
Wheat	180	£510	£ 91,800
Barley	90	£450	£ 40,500
Beans	90	£440	£ 39,600
Setaside @10%	40	(£242)	£ 9,680
Total Gross Margin	400		£181,580

Fixed Costs per hectare			
Regular Labour (paid)	£115		
Regular Labour (unpaid)	£ 30		
Casual Labour	£ 5		
Total Labour		£150	
Machinery Depreciation	£ 85		
Machinery Repairs	£ 45		
Fuel, Oil, Electricity	£ 33		
Contractors	£ 20		
Vehicle Tax and Insurance	£ 7		
		£190	
General Overheads		£ 70	
Total Fixed Costs		£410 x 400	£164,000
Net Farm Income			£ 17,580
Rent from two Cottages (net of management)			£ 12,000
Rent from storage of Caravans			£ 3,000
Net Total Income before Rent and Interest			£ 32,580

Notes:

Figures taken from 31st Edition of *Farm Management Pocketbook* by John Nix.
Rotation is typical for many Grade 3 arable farms, with two successive crops of winter wheat followed by one of winter barley and one of spring beans.
Setaside is taken directly from the amount payable for the year in question.
Unpaid Labour refers to the value of the time worked by the tenant and his family, but excludes managerial remuneration which is assumed to be part of the net income received after all other deductions.
Excludes rental value of farmhouse and of shooting rights.

remain fixed and changing trends in the markets meanwhile as discussed in Section 6.1.3 below.

6.1.2 Market Evidence

The potential difficulties that one might encounter in finding suitable comparable evidence has already been alluded to in Section 6.1 above. It will be understood from earlier discussions, principally in Chapter 4, that the market for rented farms is very limited. Whilst some land is being offered to let each year, this will of course in England and Wales be under the *Agricultural Tenancies Act* 1995 and will not be directly comparable to existing tenancies under the *Agricultural Holdings Act* 1986. Also, there is a constraint upon applying evidence of new lettings to the assessment of traditional sitting tenant rents, as is explained in Section 6.1.3 below. Therefore, for the majority of farms still let under the traditional legislation, one will need to work with examples taken from rents being paid by other such sitting tenants under their existing leases.

Unlike the sale of land which, as discussed in Chapter 3, tends to be publicized in the media and around the market place, rents are generally agreed very privately between landlord and tenant or their agents. A landlord or agent may well be willing to mention to a valuer the rents being paid on their properties and this will certainly provide some useful guidance but it cannot normally be quoted as evidence without the consent of the tenant concerned. As this may not be forthcoming, it becomes necessary to look to the national statistics instead. These originate from an annual survey conducted by DEFRA and by the National Assembly for Wales Agricultural Department and published directly by these respective government departments and also quoted in *Farmland Market* as reproduced in Figure 6.2. They do however only cover England and Wales and, although broken down into regions, do not distinguish between the different farming sectors nor the type of lease. The results are, never the less an accepted measure of general trends and averages even if they cannot provide specific comparables. To that extent they provide a useful indicator of market movements in the same way as the land price indices mentioned in Section 3.3 and 3.4.

The formal point of reference for any rental valuation is the way in which it would be assessed for review, which is outlined in Section 6.1.3 below. If a review cannot be agreed voluntarily, the matter will go either to arbitration if under the 1986 Act in England and Wales or the 1991 Act

Figure 6.2 Reproduced with permission of *FARMLAND MARKET*

Fact on rents: The Annual Survey of Tenanted Land England 2001

TABLE A – Average rent by agreement type (£/ha)

	1999	2000	2001	% change 00/99
Full Agricultural Tenancies	118.62 (± 2.11)	114.76 (± 2.57)	110.09 (± 5.16)	-4%
Farm Business Tenancies for 1 year and over	173.78 (± 4.27)	164.98 (± 3.46)	162.08 (± 5.89)	-2%
Seasonal Lets of less than 1 year	131.34 (±25.05)	123.77 (± 15.71)	134.25 (± 23.82)	8%
Full Agricultural Tenancies with provision for a rent review				
Average rent for the current year	122.34 (± 9.19)	122.54 (± 7.80)	97.59 (± 16.07)	
Average rent for the same agreements in the previous year (prior to the review)	118.24	130.83	104.68	
% change	3%	-6%	-7%	

TABLE B – Coverage by agreement type

	1999	2000	2001
Full Agricultural Tenancies			
Number of agreements in survey	5322	4471	1880
Area of land covered by survey ('000ha)	320.2	294.3	106.0
Total area of rented land ('000ha)	2320.8	2158.1	2127.9
% coverage of survey by rented area	13.8%	13.6%	5.0%
Farm Business Tenancies for 1 year and over			
Number of agreements in survey	3751	3862	1804
Area of land covered by survey ('000ha)	155.1	167.3	71.7
Total area of rented land ('000ha)	441.6	583.2	713.9
% coverage of survey by rented area	35.1%	28.7%	10.1%
Seasonal Lets of less than 1 year			
Number of agreements in survey	1421	1278	539
Area of land covered by survey ('000ha)	27.8	25.7	8.0
Total area of rented land ('000ha)	402.5	357.4	333.3
% coverage of survey by rented area	6.9%	7.2%	2.4%
Full Agricultural Tenancies with provision for a rent review in 1998			
Number of agreements in survey	543	460	284
Area of land covered by survey ('000ha)	29.7	45.8	28.2
Total area of rented land for ALL Full Agricultural Tenancies ('000ha) – see note 1.	2320.8	2158.1	2127.9
% coverage of survey by rented area	1.3%	2.1%	1.3%

in Scotland or to some similar independent process for FBTs. Whilst recourse to such measures is happily quite rare, it is none the less a common practice that when preparing a rental valuation one does so as if it had to be considered by an arbitrator. This has particular implications for the way in which market evidence might be used. In property valuations, the evidence of sales of comparable farms tends to be applied on a fairly broad basis, but if one were to have occasion to refer to another similar farm in a rental matter, it would have to incorporate a greater amount of detail. It focuses much more on the productive capability of the land and the financial value of the facilities such as buildings, rather than questions of amenity or other market attractions. It will be important therefore to take full account of those practical aspects that might hamper the farm business, even down to matters like parts of fields that tend to lie wet or old cubicle housing. In those cases, by way of illustration, the wet fields might reduce the arable potential of that part of the farm and the cubicles may be too small for today's breed of cows and cause losses through injury or infection. Similarly, there could be factors which give a farm an added advantage over others against which it is being compared, so that a higher rent would be justified at least on those counts.

6.1.3 Rent Reviews

To determine an acceptable level of rent on a farm or block of land, one needs to observe certain conventions, in the same way as property valuations are done in accordance with an established basis of definitions set out in the Red Book. In the case of agricultural rents, the reference is to the methods by which a dispute would be resolved. Under the *Agricultural Tenancies Act* 1995, both parties are free to agree whatever arrangements they wish as to how the rent should be reviewed and by what means any disagreements would be resolved. In many cases, however, these Farm Business Tenancies (FBTs) do in fact refer back to the statutory measures that have applied to all leases created prior to September 1995. The reason for this may be that the previous system has been well tried and tested over the years and is also clearly and comprehensively defined.

This refers to Section 12 and Schedule 2 of the *Agricultural Holdings Act* 1986 which sets out the factors that should be taken into account when negotiating a rent review and also those that must be excluded. It aims to establish an equitable scenario by which all cases can be assessed, incorporating the following essential features. In Scotland this comes under Section 13 and Schedule 7 of the *Agricultural Holdings (Scotland) Act* 1991.

The rent is as might be agreed between willing and prudent parties. It must take account of all relevant factors, both as to the nature of the property and the terms of the lease. Regard is to be had to the productive capacity of the holding and also to the current level of rents for other comparable farms. Scarcity value is however to be ignored, as is therefore evidence of high rents being tendered for new tenancies whether under the 1986 Act or for FBTs. In the event of a dispute, the matter would go to arbitration and the task of the arbitrator is then to interpret the situation strictly within the definitions of the Act. Disputes over FBT rents, incidentally, would normally be governed not by the *Agricultural Holdings Acts* but by the *Arbitration Act* 1996 which is slightly less well adapted to the farm situation.

Although the 1986 Act is recognized as the legal authority for farm rent reviews, there are some areas where it has been challenged. A requirement to ignore any benefit that the tenant may gain from having other land in the vicinity is in question following the case of *Childers v Anker* in 1994 and action has also been taken to try to establish that evidence of FBT tenders may be used as comparables in assessing 1986 Act rents.

Looking at the general valuation issues, however, one can see now what factors need to be considered in establishing a rent and also begin to anticipate potential problem areas. Assessing the 'relevant factors' of the holding will be a matter of inspecting the property much as in the case of a freehold valuation. In addition to this, however, one will need to investigate the terms of the lease in order to establish not only the liabilities of both parties and the financial implications arising from this for the tenant, but also to gain evidence of tenant's improvements and of redundant buildings. This raises a number of points already considered, notably in Section 4.3.

As there is a scarcity of new lettings and also difficulties over using other let farms as formal comparables, greater reliance is put on establishing the 'productive capacity' of the holding and the need therefore of working out farm budgets.

Having worked out a mutually agreeable budget, however, showing a figure of net farm income before rent and interest, one still needs to convert it into a fair rent. Generally this is done on a rough rule of thumb of dividing the net income equally between landlord and tenant. Although quite workable in practice, there is little science behind this seemingly arbitrary apportionment and it may of course need to be adjusted in the light of specific circumstances. This can arise when a tenant has invested in a larger than normal share of the fixed equipment or owns a large amount of milk quota. The significance of milk quota within rental assessments is dealt with in Section 8.1.1.

Under the 1986 Act, rents may not be reviewed more frequently than every three years. The figure that is agreed between parties must therefore hold for at least that period and it is sometimes suggested that some allowance should be made for anticipated changes in profitability meanwhile. A rental valuation is, however, made for a specific moment, just as in the case of a property valuation, and would not normally include such an element of forecasting. In preparing a budget, on the other hand, it may be appropriate to anticipate changes in such items as crop prices or interest rates as may occur during the farming year.

6.1.4 Amenity and Housing

There are some features which would enhance the freehold value of a property but which may not be reflected in a higher agricultual rent. Amenity, for example, is recognized as commanding high prices in some areas of the property market, but as it does little to improve the commercial position of a tenant it may not be accepted as a justification for setting a higher rental value too. Similarly, there are many cases where if a vacant farm for sale includes a fine house, it will attract a premium on the price, but if the property were let, the tenants might argue that the house was a burden to them in that it was an unwarranted expense against the farm business. This can be particularly so in the case of many traditional leases whereby the tenant is required to reside personally in the farmhouse. Whilst this may have been a sensible management criterion in the past, it would now prevent tenants from sub-letting their houses and thereby gaining a proper financial return from them. There are no clear-cut answers to such situations which tend to be resolved more by negotiation according to the circumstances rather than by any hard and fast rule.

It has already been mentioned (in Section 2.2.2) that farms may well have cottages that are no longer needed for agricultural workers. These might be taken back in hand by the landlords, but frequently they remain within the lease so that the farm tenants would be entitled to let them out on shorthold tenancies. In many areas the rents obtained from such lettings can represent a significant income and would certainly be taken into account when assessing the rental value of the farm itself. Indeed, research by FPD Savills shows that in 2002 as much as 40 per cent of income on estates in the higher value areas was derived from residential rents.

6.2 Valuation of Tenancies

There are occasions when agricultural tenancies may need to be valued separately from the property over which they are held. This can be for the assessment of tax liabilities or for compensation following compulsory purchase or as a basis for negotiation between the landlord and tenant or some other interested party. However, most farm tenancies are non-assignable and are in the names of individual people rather than companies, which means that they cannot normally be traded and that there is then virtually no open market evidence upon which to base one's assessment.

Such valuations are therefore necessarily rather theoretical and will tend to follow one of two main alternatives, namely either to consider the differential between the value of the property when vacant and as let or to capitalize the income that the tenant gains from his right to occupation.

The first of these methods should present little difficulty in itself because there are, as described previously, straightforward methods by which the property may be valued as if with vacant possession and then as subject to the relevant tenancy. The differential, or vacant possession premium, will be determined simply by subtracting the tenanted value from the full in-hand figure. The question then arises, however, as to whether it is fair to assume that this really represents the price that someone would pay to acquire the right to rent the land should there have been an open market opportunity to do so. After all, the fact that the tenanted market trades at a lower level than that for vacant land is a reflection of demand on the part of investors and of the rates of return that they might accept on their capital, rather than a measure of the tenant's interest. The rationale behind using this calculation depends more upon the fact that due to the security of tenure, it is the tenants who are more likely to have the chance to realize the vacant possession premium, and that this therefore equates to the value of their lease. Landlords would normally only be able to gain this premium in full upon the death of a tenant whose heirs had no claim to succession.

In practice of course it is unlikely that a tenant would be able to benefit from the full amount of the premium as landlords will seek to negotiate a share in it by, for example, selling the freehold to them at more than the actual tenanted value. The amount by which this is shared will depend on individual circumstances, including the age of the tenant and the likelihood of an heir claiming succession as well as the value of his fixtures and capital improvements and the arrangements for quota. The accepted rule of thumb is to assume that initially the premium might be shared equally between the two parties and then adjusted according to any special circumstances as may apply.

Valuing a tenancy by deducing the vacant possession premium has the advantage that it uses familiar market practices and seems appropriate for situations where landlord and tenant are intending to dispose of their interests. It may be less suitable, however, for deriving a figure to represent the worth to the tenant of having an agricultural lease as the basis of his business and for his home. The alternative and more commercial approach is then to consider the capital value to the tenant of being able to carry out his business by virtue of his lease on the property. The concept of capitalizing a profit rent is not unusual for certain forms of investment, but it does need careful qualification when applied to agricultural property.

One practice that is used, particularly in compensation cases, is that where the present rent is likely to be increased at the next review, then the amount by which it would rise represents the profit rent and is the advantage for which the tenant could, theoretically at least, pay a capital sum. As rents on statutory leases may be reviewed every three years, this will tend to come to a relatively insignificant amount since the rent will probably already be close to a current market level, having been reviewed only a year or so previously, and since the capitalization will be for an equally short time too. The theory is then sometimes extended with the suggestion that there can be a further profit rent on the grounds that tender rents are above those that are reviewed under statutory regulations. There can be difficulties, however, in using the evidence of tender rents as a measure of farm profitability. Such rents may be justified by the prospective tenants only on special and short-term grounds. A lack of farms to let, an opportunity to gain economies of scale and the knowledge that any excess rent agreed at the present time could be reduced by statutory negotiation or arbitration after three years, can all create a distorted position. The case of new lettings under the 1995 legislation is dealt with in the next section, but it might be noted here that the possibility of a downward review in the short term is likely to be precluded by the initial terms.

The capitalization of a profit rent may seem a little theoretical but it has been used by District Valuers as became clear in the case of *Walton*

v. Inland Revenue Commissioners (1994 & 1996). The calcula
would follow the principles outlined in Chapter 5 and can be illustrate
as follows:

Assumed current rack rent per hectare		£ 150	
Present rent last reviewed 18 months ago		£ 125	
Current profit rent		£ 25	
YP @ 5% for 1.5 years		1.41	
			£ 35.25
Assumed tender rent per hectare	£ 270		
Rack rent	£ 150		
Additional profit rent	£ 120		
YP @ 12% for 3 years	2.40		
		£288.00	
Value of Tenancy		£323.25 per hectare	

The above in fact illustrates the conclusions reached by the Tribunal in the Walton case, which has been more fully analysed in *The Valuation of Agricultural Tenancies: Art or Artifice* by Charles Cowap. Much of this is outside the normal field of practice and therefore beyond the scope of this book, but in summary it could just be noted that the use of the 12 per cent rate was in recognition of the commercial risk attached to a high tender rent and that the period of 3 years was probably chosen as being the time that would elapse before a new tenant would be able to seek a rent review. The earlier figure of 1.5 years is taken in this example as being the average time before review in an existing statutory tenancy.

In one of the rare cases when a tenancy becomes available for purchase, such as under an assignment or through the sale of a company, the profit rent may be taken as the difference between the full amount of anticipated net income to be earned by the tenants from the farm and the rent payable during the foreseeable future. In some instances, new lettings might be made on the basis of an agreed initial rent for a certain term together with a capital payment at the outset. This payment may be in respect of certain tenants' improvements that are to be taken over but can also be in part an equivalent of capital value.

6.3 Lettings under the *Agricultural Tenancies Act* 1995

Tenancies created under this more recent legislation have tended to be for shorter terms, without on-going security of tenure, at high initial ed provisions for reviews and on a non-assignable circumstances there is little reason to attribute any value to the tenancy as the profit rent is reduced to an mercial level and the right to occupy the land is for a .

however cases of Farm Business Tenancies being agreed for longer terms and where any break clause is linked to a formula of compensation to the tenant. There are also instances where these longer agreements are set at relatively modest initial rents but where the incomer is required to pay a capital sum on entry. This may be ostensibly for the value of specified fixtures and fittings but it may represent an element of 'key money' as well. In these circumstances it might be that the tenancy will be considered to have a value and that this would be assessed using the same principles as for statutory leases, namely in identifying and quantifying the financial advantage accruing to the tenant.

Check-list – Rental Valuations:

- Reference to arbitration rules under the *Agricultural Holdings Act* 1986.
- Need to assess both property features and productive capacity.
- Limitations on finding and using comparable evidence for 1986 Act tenancies.
- Calculation of farm budgets, using Gross Margins and management data and tables.
- Conventional 50:50 split of net income between landlord and tenant, with variations according to circumstances.
- Allowance for time factor and non-farming features.

Valuation of Tenancies:

- As equivalent to all or part of VP premium, or as capitalization of profit rent.
- Reliance on theoretical approach, and lack of market evidence.
- FBTs and lesser capital values due to shorter terms and/or higher rents.

CHAPTER 7

Land Let on Farm Business Tenancy

Lettings made in England and Wales after 1 September 1995 are subject to the *Agricultural Tenancies Act* and will be in the form of Farm Business Tenancies (FBTs). As mentioned in Section 1.1, the essential distinction between these two forms is that the latter can be for a specific fixed term at the end of which the landlord is legally assured of being able to regain possession. When valuing land let on this basis, it would seem logical then to use the principle illustrated in sub-section (vi) of Chapter 5 of applying a Yearly Purchase (YP) to the remaining rental period and a deferment to the ultimate vacant possession value. There are, however, circumstances where this may not necessarily produce a totally realistic answer.

The majority of FBTs seem to be arranged on relatively short terms of around three or five years and often on bare land rather than on complete farms equipped with houses and buildings. Their purpose is often to provide the tenants with additional land over which they may spread their existing overheads and thereby gain some economies of scale. This in turn allows them to pay an enhanced rent for the marginal land as it will still leave them in a more profitable position than if they had been unable to expand their operations. Furthermore, such tenancies are likely to be secured only against competition, since there may be more than one farmer seeking extra land and needing to have it in a specific locality that lies within an accessible distance of the existing holdings. Also, since the tenants in these cases have an existing enterprise that may be owned or held on a secure statutory lease, they can accept the risk of farming the additional land on a short-term basis. Even on the rarer occasions when whole farms are let on longer terms of 10 or 15 years, prospective tenants may be forced to offer a premium rent in order to be able to secure the opportunity. The reasoning behind the more common practice of using relatively short terms is partly that the owners will feel that they avoid the problem of devaluing the land as would have happened with the traditional statutory leases when creating a lifetime tenancy lowered the value of in-hand land by 30 or 40 per cent or more. Such reasoning may however not be borne out in practice.

At the moment when the new letting is implemented the land will have been at full vacant possession value, determined by the interaction of buyers and sellers interested in farming it in hand. The price may therefore be influenced, and inflated, by factors of marriage value and location, as mentioned in Section 2.2.4, which do not apply to most cases of tenanted land. If the property were then to be offered for sale again while the FBT still had perhaps two years to run, one would need to rely on this same degree of interest from the market in order to achieve the equivalent of the full vacant possession figure. It is unlikely, however, that potential buyers would pay as much for land that was subject to someone else's occupation for another two years as they would for land that they could take in hand straight away. Even with a full rent being paid during the intervening time, it could be more difficult for a farmer purchaser to secure the necessary finance and he would not be gaining any advantage of economies of scale until he was able to take possession. On the other hand, it is also unlikely that a property with only a short lease would be particularly appropriate to investors, other than possibly at a significant discount.

Under a traditional tenancy the land is bought at a discount against the full vacant possession value, allowing an acceptable return on capital from the rent as well as the long-term cover of the possibility of a substantial reversion should the tenant leave. Whilst FBT rents tend to be higher than those for existing statutory lettings, there can be some risk as to whether these premium levels will be sustained thoughout the term of the lease, where it may have been agreed for a longer period. Such risk would normally be reflected in a higher yield and therefore a lower price. As yet, however, there is little guidance as to what yields would be applied in the market place and there is a danger of valuation calculations becoming rather theoretical. This predicament can be illustrated by adapting the example used in Chapter 5.

If the statutory rent in that earlier case was £15,000, one may assume that under a new FBT it might be let for perhaps one and a half times that figure or £22,500. If the unexpired term is only two years, the supposed risk of this higher level will not be as significant as in the case when it has still eight years to run. In the former instance one might not need to increase the yield by more than a nominal amount, but over eight years with presumably a review date within that time, the yield would be slightly higher again than in the original example if it were thought that there was a chance that it might then be negotiated downwards.

Rent	£ 22,500	
Less: management fees @ 3%	£ 675	
Net current annual income	£ 21,825	
YP @ 4.5% for 2 years	1.87	
		£40,813
Present vacant possession value	£ 550,000	
PV of £1 in 2 years @ 4.5%	0.916	
		£503,800
Capital value		£544,613

Even after allowing a margin of 0.5 per cent on the yield to reflect the remaining two-year occupancy by the tenant, the calculation produces an equivalent of vacant possession value which may however not reflect the attitude of potential purchasers in practice. On the other hand, from a pure investment valuation viewpoint, the deferment of only two years until possession and the relative lack of risk to the rent in that time would not really warrant a substantially higher yield than that used on a traditionally let farm. Furthermore if, as many people might wish to believe, an in-hand farm could be let on an FBT without any significant diminution in vacant possession value, then this would be the arithmetic justification for this. In practice though the market would seem unlikely to support such a position.

When the FBT might have a slightly longer period still to run, as in the example in sub-section (vi) of Chapter 5 above with eight years unexpired, there could be a relatively higher risk as regards the rental return to the investor. As has already been mentioned, FBT rents have tended initially to be much higher than those for existing statutory lettings. If this initial rent is fixed for the full term of the lease, the landlord may feel disadvantaged in that he would miss out on any inflationary growth that might arise even though the starting rent was already at a premium. Alternatively, if the agricultural economy does not grow during the period of the tenancy and farm incomes decline, then there could be a possibility that the tenant is unable to sustain his original offer and is forced to give up the land. Whilst the owner would not necessarily suffer any direct capital loss from this as he would have regained full vacant possession value, it may not suit him then to sell the property in order to realize this capital position.

The alternative would be to relet it, but presumably at a difficult time when farmers might be unable to offer the same level of rent as the failed outgoing tenant. If he were instead to take the land in hand and arrange to have it farmed directly for him, he might need to find all or part of the annual working capital depending on the type of farming arrangements available to him, which would again put him at a financial disadvantage. All these concerns would appear to justify using a higher investment yield, but it is again difficult to quantify by exactly how much.

As a general principle, it is fair to assume that potential purchasers would be less attracted to an FBT investment involving the longer commitment of eight years and would therefore need to see a valuation figure below that applied to the farm when let for only a further two years. There is insufficient market evidence from which to deduce the appropriate level of yield that such investors would accept, but it could be argued that there is little difference between an investment let on an eight-year FBT and the reversionary interest shown in sub-section (vi) of Chapter 5, other than that the former is let at a substantially higher rent. This higher rent may represent a better return but it does also involve a degree of risk in whether it can be sustained over the remaining term. As a basic investment principle that risk will need to be reflected in a correspondingly higher yield. If that were to be taken at 5.5 per cent instead of 4.5 per cent in the previous example, it produces a very similar capital value to that of the reversionary interest:

Rent	£ 22,500	
less: management fees @ 3%	£ 675	
Net current income	£ 21,825	
YP @ 5.5% for 8 years	6.33	
		£138,152
Present vacant possession value	£ 550,000	
PV of £1 in 8 years @ 5.5%	0.652	
		£358,600
Capital Value		£496,752

If, however, the level of the FBT rent is not perceived as being a significant risk, then investors might well pay a correspondingly higher figure by accepting a lesser yield. It might be also that one would question whether the rate applied to the deferment should be higher than that for the initial YP, allowing for the element of uncertainty over the eight-year period and as discussed in example (vii) and illustrated in example (viii) of Chapter 5. In fact, with a properly executed FBT there should be little risk that possession might not be obtained on termination and therefore little justification for a higher yield on the PV calculation.

One may not be able to produce a definitive theoretical solution in these situations, but they do highlight the difficulties that need to be addressed when working without adequate real market evidence. In the more customary cases of traditionally let or vacant land, the valuer has to use an increasing degree of judgement when there is a shortage of evidence, as mentioned in Chapters 2 and 3. A similar situation arises in this relatively early stage of the existence of FBT.

A number of other points of practice are illustrated in these workings of valuations of properties let on FBTs.

First, the management fee of three per cent has been estimated at a lesser rate than that used for the equivalent statutory tenancy on modern terms. This is because the FBT rent is considerably higher than that on an existing traditional lease and it emphasizes that such percentage rates are sometimes only approximations of the kind of terms that might be agreed in individual cases. Ultimately, the fee agreed for estate management will relate to the assumed amount of professional time that is likely to be spent on that function each year. A short-term FBT arranged on a basis involving the landlord in little regular maintenance may produce perhaps one and a half-times the rent of a traditional lease, but it would hardly involve the agent in one and a half-times the amount of professional work or responsibility. By adopting a figure of three per cent, as opposed to five per cent, the valuer has made an allowance for this situation as it might arise in practice. As a secondary issue, one may also be aware that short-term leases will need reletting or renegotiating on a relatively frequent basis and that the agent's annual remuneration might be intended to cover the additional work involved in this.

Mention has been made earlier of the fact that rents agreed for FBTs tend to be higher than those for existing statutory leases and although the main valuation implications have been explained, one needs also to understand some of the reasons behind this. Even under the *Agricultural Holdings Acts*, new lettings would tend to command a premium over existing sitting tenant rents. This was due to the shortage of such tenancies becoming available, to the relative prosperity of farming during much of the recent past, to the system of rent reviews and to the valuation and taxation implications for landowners. The shortage of new lettings meant that there would be keen competition for any that did become available and that prospective tenants would be encouraged to bid high figures in order to secure the farm. They could in most cases justify this on the strength of recent and anticipated income and also knowing that should circumstances change, they would be able to have the rent reviewed after three years and then probably reduced to a level more comparable to existing tenants and therefore effectively removing the initial premium. Rent reviews were governed by statute and were required to exclude the evidence of anything other than sitting tenant rents. The possibility of a downward review has been mentioned in Chapter 6.

That new lettings were so scarce was due to the fact that tenants had such long terms of tenure and therefore rarely vacated a farm voluntarily, particularly as there was unlikely to be another let farm to move to. When a vacancy did occur, however, landlords would be disinclined to relet as it meant tying up the land for another generation or longer and reducing its value from the full vacant possession figure that would apply at the moment that it came in hand. They would also have been disadvantaged on both income and capital tax were they to relet the holding. The *Agricultural Tenancies Act* 1995 sought to redress many of these constraints through the introduction of FBTs, but as yet some of the distortions remain. In particular there has still been a shortage of land being offered to let and such lettings as are being made available have been for bare land rather than complete farms and have been only for short terms of only three or five years. Meanwhile, prospective tenants have, as mentioned earlier, felt justified as well as compelled to bid high figures in order to gain the additional land over which to spread their existing costs, knowing too that the commitment was being made for just a short period. As a result, new FBTs tend to be at premium rents, dependent on economies of scale and on prevailing good conditions and are often for bare land holdings that are less attractive to investors than fully equipped farms. One can see then that there is some difficulty in reconciling all this when deriving investment valuations.

Whilst it may seem from this perspective that the latest legislation has brought little change to the market, there are some features that have made a fundamental difference. Land let on FBTs was granted the same capital taxation relief as in-hand land which should make it more attractive to private investors. FBTs are also no longer constrained by statutory regulations on rent reviews and it is therefore possible to agree an arrangement which avoids the possibility that a high tendered rent might be reduced after an initial period. This should remove some of the uncertainty about longer-term returns from newly let land, although it should be remembered that the majority of FBTs are only for short periods. This means that the landlord could then still face the risk that the tenant will not renew his lease after only a short time other than at a reduced rent and that other potential lessees would also only offer lower figures according to the then prevailing agricultural conditions. Tenants are less able to agree any longer terms unless they contain a break clause, which effectively reintroduces the element of uncertainty for the landlords. Landowners, on the other hand, may anyway be unwilling to let out land for longer periods as the valuation problems then become more evident.

These valuation difficulties arise ultimately from the fact that when farmland is let it switches from one market to another and in the case of short-term FBTs it may even fall uneasily between the two. At the moment that it is available for letting, the property would still be with full vacant possession and as such its value would be derived from the competitive interest of purchasers looking for in-hand land. Many of these are likely to be able to benefit from gaining special returns from farming themselves often in conjunction with an existing holding. Once it is let, even on a short term, it would be of less interest to such buyers but would also command only limited demand from alternative purchasers such as investors. With statutory tenancies this was easily recognized and the relative valuations reflected this clearly with a substantial differential. Under FBTs it is assumed that this discount will not apply when the letting is only for a relatively short term, but it will none the less arise, as illustrated above, and the valuer will need to take all this into account.

Check-list – Land let on Farm Business Tenancy:

- Question of discount on land let on FBTs – assumed to be negligible for short-term agreements, or for those with break clauses.
- Use of conventional investment approach for longer-term lettings.
- Need to check theory against practice and against market assumptions.
- Allowance for reviews on initial high rents.

CHAPTER 8

Quotas, Licences and Designations

Whether farmland is vacant or let, its value will depend in some measure on the income that the farm is able to produce. The degree of influence will of course vary according to the type of property and will be more evident in the case of, for example, a commercial block of bare land than for an attractive residential farm with a fine house and other amenities. This income is in turn dependent on a whole range of factors, many of which have already been considered in earlier chapters. They include the essential physical features of the farm such as the nature of the land and the quality of the buildings and also its location and layout. There are however other matters that can be crucial to a farm's profitability and furthermore have a direct or indirect effect on value.

Farming in Britain is closely controlled by European and other policies, whether through financial mechanisms or by other regulatory means. In the majority of cases, individual farm incomes will be determined by the eligibility or otherwise of that farm qualifying for aid or alternatively suffering restrictions. Some of these will stem from registrations made in the past and which cannot now normally be added to or removed. As these rights or constraints have such an influence on income, the fact whether the farm was properly registered at the appropriate time will have an ongoing impact on value. This issue can best be demonstrated by considering the main current examples.

8.1 Quotas

Quotas have been introduced under the Common Agricultural Policy in a number of sectors with the purpose either of restricting total production or of supporting individual incomes or both. Most of these are linked directly to the product and therefore to the producer rather than to the property itself, but in one case, namely the dairy sector, the quota has actually to be tied in to a particular area of land. Each type of quota can have an impact on land value, but one should start by first looking at milk quotas.

8.1.1 Milk Quotas

In April 1984 a system was introduced whereby dairy farmers throughout the European Community were given an entitlement to produce a certain quantity of milk each year. In Britain this is governed by the *Dairy Produce Quotas Regulations*. The entitlement was granted in the form of a quota specifying the permitted annual volume of production for each farm. The defined level of production was assessed according to the circumstances of the holding at that time. From then on any milk produced within that annual maximum could be sold in the existing national markets as before. If however a farmer exceeded his allocated quota before the end of a year, he would be liable to pay a penalty in the form of a levy. Whether this levy is demanded by the administrative body, which is the Rural Payments Agency, and the rate at which it is charged, will depend on whether the country as a whole has exceeded the national quota. These two circumstances can however easily coincide as, for example, a good forage year will result in increased milk production across the board. Farmers without quota would have had to pay levy on all their production and were therefore effectively barred from dairying other than for a few specialist markets.

One can see that from the very outset the existence of milk quota on a farm would have a vital impact on property values, even though initially the quota had no tradable value in itself. There would be many farms that could be viable only as a dairy unit, due to size, location and terrain and to the investment in specialist equipment such as the milking parlour. Without quota those holdings would be restricted to other enterprises such as suckler herds or sheep which would not fit into the scale of a dairy farm nor provide a similar level of income. If unable to generate an adequate income, the value would be significantly reduced against that of the equivalent farm trading under a quota.

This situation is however now more complicated due to the fact that, contrary to the original intentions of the European Commission, milk quotas have themselves become valuable. If a farmer was likely to produce more milk than for which he had quota, he had three options: to pay the levy, to throw away the surplus milk, or to acquire an additional amount of quota to cover the excess. Inevitably, means were found whereby quota could be bought or sold or even leased, whilst still being technically tied to a particular parcel of land. Circumstances were that such quota could command high prices, even though there was no guarantee that the system would continue for more than a few years. Prices have in fact been at a level that quota can be worth over half as much as the farm itself. There have been times when quota was traded at 50p per litre and more, which for a herd of 100 cows averaging 5000 litres per annum represents £250,000. Such a herd might be sustained by a farm of 150 acres, with a value then of around £400,000 depending on circumstances. This occurred due to the ongoing profitability of dairying and the relatively high financial cost of levy, but is also a reflection of the fact that most quota is bought or leased in marginal amounts rather than for a full farm production. To avoid levy, one needs to acquire just enough quota to cover the anticipated surplus and can afford perhaps to pay almost as much as the cost of the levy itself. Also, anyone expanding their herd or the output per cow beyond that which applied in 1984 would again be buying in quota at a marginal amount and able to spread the cost over a larger and more profitable business.

Whilst the quota market may be made up of buyers acquiring in the main only small lots at premium prices, the vendors could well be dairy farmers selling their herds and leaving the industry. Anyone owning quota must therefore reckon with it all being worth the full price even if sold in small lots. Also, there are occasions when the purchaser of a farm will have to buy in the total required quota at the going rate, whether off the vendor of the farm or through the market.

When valuing a dairy farm, one must therefore allow for the full cost of quota. In an actual property sale, this may often in fact be kept as a separate item and a valuer can therefore be justified in keeping the two figures separate as well. It can then be a question of whether there is more evidence of dairy farms being sold inclusive of quota or whether the valuer feels more confident of deriving a figure for such farms exclusive of quota and then adding on the appropriate volume to suit the stock capacity of the holding at the then current price. One should be aware, however, that in practice when a farm is sold without quota, it can be at a slight discount to what would be reckoned as the sum of the net property value and the prevailing price of quota. Even though the necessary amount of quota can be bought in the open market, buyers may have some concerns about making such a substantial cash investment in this form, particularly at a time when there may be increasing uncertainty about the long-term future of the entire system.

In the case of a let farm, there is a further issue in that the tenant is also likely to have an interest in the milk quota. This may affect the assessment of net land value as described immediately above and will also be of consequence in determining the correct rental income for the holding and the compensation that would be due at the end of the lease. Tenants would not have been granted ownership of quota under the 1984 Regulations but they may meanwhile be entitled to claim some in their own name. This could arise from one of three principal reasons: either that the tenant had acquired it in his own right; or that a substantial proportion of the fixed equipment on the farm has been provided by the tenant; or because it can be established that the tenant is producing an above-average quantity of milk.

In the first of these cases, the amount of quota acquired by the tenant will have been formally registered with the controlling body, now administered by the Rural Payments Agency. In the latter two instances, the amount can be established by references to a process laid down in the *Agriculture Act* 1986. This defines for a wide range of dairy farming situations a Standard Quota which can be related both to investment in capital items, such as parlour or dairy equipment, and to the level of actual production as indicated by the quota allocated in practice to that holding. Where high production levels are deemed to arise essentially from the way in which the farm is being run, it is the tenant who becomes entitled to what is known as Excess Quota, which is the amount by which the official Allocated Quota exceeds Standard Quota. Where it is a matter of capital equipment provided in the past, there is provision for Standard Quota to be apportioned between each of the parties according to their respective investment. From this is then derived the 'tenant's fraction', which quantifies the volume of quota due in respect of that investment.

The prevailing price of quota per litre can be obtained either from *Farmers Weekly* or by reference to an agent specializing in that market. Allowance may have to be made for possible temporary distortions arising from the fact that milk quotas are administered on a yearly basis ending on 31 March and also for adjustments for differing butterfat content. In the example of a full property valuation given in Figure 2.1 in Chapter 2, it will be noted that the quota is described not only in terms of volume allocated but also as to the percentage of butterfat and the proportion already used up at the time. A more detailed account of these various arrangements can be found in *Milk Quotas Explained* by John Edwards.

8.1.2 Livestock Quotas

A quota regime also exists for sheep and beef, but these differ significantly from milk quotas and are of less consequence to property valuations. Whilst milk quotas are an essential necessity in dairy production and are attached to land, these other livestock quotas are held by the producer and not linked to any particular property. They are also not coupled with a penalty system of levies but are instead a means of gaining a premium over ordinary market prices for sheep and beef, provided that certain conditions have been met. It would therefore still be possible to produce sheep and beef without quota, albeit less profitably. As the landowner has no involvement in these quotas, the implications for rent and compensation that can arise in the dairy sector do not apply and there is no direct effect on property values. An indirect influence can occur, however, in that to qualify under the present system of quotas the farmer has to adopt a progressive policy of extensification. This means that he will receive the premium only if he reduces the number of stock carried per hectare during a prescribed period. To benefit from the additional income he will therefore either have to reduce the number of animals on his farm, and thereby reduce the basis of his income, or find additional land on which then to graze his existing stock. To achieve this latter option, he may either rent land if available or otherwise try to buy it. As many farmers in one area will be seeking to do the same, the local land market comes under pressure and prices are likely to rise with the ensuing consequence on valuations.

8.1.3 Arable Quotas

Arable quotas as such no longer exist, following the deregulation of the Potato Marketing Board, although sugar beet is still controlled by contracts which are defined in terms of different bands of quota. These contracts are awarded to growers and do not run with the land, although the ability to produce sugar beet can have an influence on land value and on rents. Sugar beet can be grown successfully only on certain types of soil and in a location that is within a reasonable haulage distance from a processing factory. Such land that can therefore benefit from a sugar beet contract may command a premium. As the crop is vulnerable to infestation and has to be grown on a long rotation, growers with a contract may hire land off neighbours on a seasonal basis in order to rest their own land. The financial advantage of having the physical capability of growing sugar beet does thereby spread even to land owned or occupied by people who do not have a contract. The contracts are controlled by British Sugar and are currently awarded without cost and have no tradable value. This system is, however, under review and there could be changes which would have an impact on valuation. Meanwhile, the amount by which the value of relevant land may be enhanced can only be determined by reference to general local market evidence rather than any specific analysis or calculation.

Arable land is also currently regulated by a system of Aid Payments, as outlined in Section 8.3.1.

8.2 Licences

Irrigation Licences can have an effect on land values. As was mentioned in the section dealing with Land Classification, in Section 4.3.6, it is the lighter and more workable soils that have the capability of growing many of the more specialized crops such as roots, bulbs and salads. Such soils allow for better sowing and easier harvesting, even under wet weather conditions when it would be difficult to do so on heavier land without damaging the crops or machines. It will have a tendency to dry out, however, and successful cropping will depend on ability to irrigate the land. The source of irrigation may be above ground in the form of watercourses or specially constructed lagoons or from bore-holes tapping resources under the property. In most similar cases, this water may only be used under a licence from the Environment Agency which will specify the annual volume of water that may be taken, the times of year when this may be done and the fee that is payable. A valuer must check that these conditions are adequate for the farm and that the property is not handicapped by not being able to draw enough water or do so at the time of year when it might be needed. One must also consider what effect there might be on the profitability of the farm if the licence were reduced or removed at some stage in the future and the likelihood of this ever occurring. With apparent climatic changes and with growing populations taking an increasing amount of local water resources, it is becoming more possible that such restrictions might be imposed.

The premium applicable to fertile light land with irrigation will again be a matter of analysing local market evidence, but the discount for having a licence revoked or reduced will be a matter of assessing the change in farm income, as in the example of grain storage in Section 2.2.3, and of the circumstances that would determine whether water resources are vulnerable in that particular locality.

8.3 Designations

Controls and regulations affecting the countryside are developing constantly and it would be impractical to try to provide a list of each designation that might currently influence land values. It is however important to be aware of the kind of factors that need to be considered and taken into account. Such designations may be defined loosely as either voluntary or compulsory and also as either restrictive or supportive.

A 'voluntary' designation is one where the farmer or landowner has the choice as to whether he will participate in a scheme operating within the designated area. An example of this would be the Environmentally Sensitive Areas or ESAs which have been applied to specific localities or landscapes and within which grant aid is available to those farmers who enter into management agreements. These agreements limit certain commercial agricultural practices and favour less intensive or more environmentally friendly systems instead. The grants are intended as compensation for any reduction in income that may arise from adopting these practices. There should therefore be little financial difference between adopting the scheme or farming conventionally.

A 'compulsory' designation, however, such as a Site of Special Scientific Interest or SSSI, is imposed by regulation and does not normally benefit from any financial compensation. Some financial recompense may be payable by the appropriate regulatory body, such as English Nature in England, in conjunction with a management agreement and based on an estimate of profits foregone. However even with such compensation, a farmer or landowner could still be disadvantaged if the designation were to restrict his normal commercial activities on the land such as the application of agrochemicals.

There are arrangements that are restrictive on the occupier as in the case of a property lying within an Area of Outstanding Natural Beauty (AONB) where there are likely to be, for example, tight building controls. Some designations are however supportive in that they entitle the farmers to special subsidies, as arises with Less Favoured Areas (LFAs) and the Hill Farm Allowances (HFAs) in England or the Tir Mynydd Scheme in Wales.

How these various categories affect land values is difficult to judge, largely because there is so little actual market evidence. For example, if a property lies within an LFA the occupier will be entitled to HFAs and therefore have an additional form of income not available to an immediate neighbour whose land may lie just outside the designated area. If two such farms came up for sale, would the one in the LFA command a higher price because of the additional income or would the market consider that such grant aid is genuinely only a form of compensation for the greater hardship endured in such Areas? It may depend on whether the boundaries of the LFAs are drawn sufficiently accurately as to include only those farms that are disadvantaged. In the event, there is no real market evidence from which to deduce any ruling on this and the valuer can only take care to have noted the existence of the designation and to anticipate any possible implications that might arise from it. For example in the case of an LFA, there may be concern over whether the HFA payments will be maintained in the future at their previous levels. Any doubt over this represents a risk to the farm's profitability and therefore ultimately to its value too.

The effect that a compulsory and restrictive designation such as an SSSI may have on value is probably confined just to the area over which the designation applies. In most instances, this will be a relatively limited part of the whole property, such as a copse or wetland meadow, and so not materially affect the viability of the farm. The area itself could then be assessed at a nominal figure per hectare and added like one of the component parts mentioned in Chapter 2.2 to the ordinary commercial figure deduced for the main part of the holding. It remains to be seen whether there will be any reaction within the land market to the present extension of Nitrate Vulnerable Zones across much of England and Wales.

For the larger, voluntary designations such as ESAs there is again no firm rule, although a number of points may arise. As there is an entirely free choice as to whether to enter into the management agreement or to continue farming without any commercial constraints, one might expect that the value of the land will be determined by the latter option and therefore not be affected by the designation. It might be questioned, however, whether such voluntary arrangements may not one day become compulsory as the importance of the landscape and the feasibility of farming it under certain restrictions has already been established. If that occurred, there would be less need for the grant payments to be maintained at their current levels when they still need to be high enough to attract farmers to take up the scheme. Under those circumstances, value would be reduced and, if this was seen to be a future eventuality, an adjustment to the present assessment might have to be made.

When management agreements were made in the past, but not maintained for their full term, there could be a liability to repay grants that had been received. In the case of a farm being bought by someone who intends to revoke the agreement made originally by the vendor, the buyer might feel the need to adjust his price to cover the cost of repaying the grant. Therefore, in a valuation of a property that had benefited substantially from such grants, some allowance may have to be made for this contingency. This is a relatively subjective matter, dependent on the circumstances of the purchaser, and for which there is little actual evidence. Such issues might be more noticeable in a weak market than under buoyant conditions.

8.3.1 Arable Aid Payments

In 1993 there was a change in emphasis in the arable regime in the European Union with payments being made more on the basis of the land used for such production rather than by supporting the price of the crops themselves. To be eligible for such aid under the Integrated Administration and Control System (IACS), it became necessary to establish that the land had been in eligible arable cropping as at the end of 1991. Where this was the case, a subsequent annual application form would entitle the occupier to aid payments that were initially at around £250 per hectare for cereals and almost twice that level for oilseeds. As this was equivalent to about 50 per cent of the Gross Margin of such crops, the payments made a substantial difference to farm income and, being attached to the land, to the value of the property too. Land that was properly registered under IACS would therefore command a premium over other land of arable quality that was unable to benefit from these payments. While this system remains in place, a valid registration will be a consideration in land value, although the amount may vary in time if the level of aid reduces. Where it cannot be established that land was in arable production at the end of 1991, it may have to be valued at a discount to reflect the lesser income opportunity, possibly at the equivalent of pasture land.

8.3.2 Set-aside

This arable regime includes the requirement that a predetermined percentage of the arable area is left uncultivated each year as 'set-aside' for which an annual payment is also made. Introduced originally in 1992 at a the level of 15 per cent, this has declined meanwhile and represents a lesser proportion of farm income. Set-aside predates IACS and could be applied for on the basis of current cropping data if the historic records are not available. It does not therefore have the same implications for value as IACS registration. Set-aside is commonly on a rotational basis, being moved to different fields around the farm, but can also be agreed on a permanent basis which could affect valuations in a similar manner to the repayment of grants as considered in 8.3 above.

Checklist – Quotas, Licences and Designations:

- Quotas when attached to the property and having separate value.
- Quotas and contracts granted to or acquired by the producer.
- Land values dependent on irrigation licences.
- Effect of Designations, whether voluntary or compulsory, and whether restrictive or enabling (for receiving grant).
- Significance of IACS registration.

CHAPTER 9

Development and Diversification

While the emphasis of the valuation appraisal has so far been on the agricultural detail, it is also necessary to assess whether all or part of the land could be used for other purposes, particularly those that would involve different markets and therefore different sets of values. This is determined firstly by whether planning consent might be granted for such alternative purposes, which in turn raises a number of points that will need to be considered:

1) Whether any planning consents have already been granted on the property but not been implemented;
2) Whether any relevant developments already carried out on the land do in fact have the necessary consents;
3) Whether it is possible that permission would be granted for any new and as yet unconsidered developments;
4) Whether there is any likelihood of planning changes elsewhere in the locality that could affect the property, either adversely or to advantage.

9.1 Planning

9.1.1 Development value

In the first case, where a valid consent exists for a purpose other than the existing use, one will expect there to be an enhancement in value over the ordinary agricultural market levels. If there is an opportunity to carry out, for example, a building development, then that area available for this purpose will attract interest from purchasers who are competent to exploit this opportunity and then to benefit from the returns achieved by it. As such returns tend to be considerably higher than those from farming, the land will justify a higher price accordingly. The valuation of that part of the property will then no longer depend on the agricultural market features, but on those that determine that particular building sector. The principles for this are broadly the same but the salient factors, and the resultant figures, will differ.

Take as an example a small field on a farm on which there is planning consent for residential development. The immediate issues that will influence the price that purchasers such as builders would pay will be those affecting the final return. These include the number of houses permitted on the site and the selling price of such houses in the current local market. Also of relevance will be the question whether extra costs are likely to be incurred due either to physical features of the land such as gradient, poor access or drainage, or to conditions that may have been imposed in the granting of the consent involving perhaps the provision of some public amenity like road improvements. For that particular part of the farm on which the non-agricultural use is permitted, the valuer will have to quantify those development factors which are likely to override the normal agricultural features that have been considered hitherto.

The existence of a valid consent for some non-agricultural purpose might lead one to assume that there will inevitably be some increase in value. This is not always so, however, and the valuer must be prepared to identify any possible shortcomings. The fact that a landowner has gained planning consent but then not implemented it might be a sign that the development was not ultimately viable. There have indeed been many examples of this where, for example, permission may have been granted for converting farmland to a golf course, when this then proved to be financially unsustainable the land remained in agricultural production and was valued accordingly. One can establish this either by assessing the potential demand from developers for that particular site or by calculating the cost of conversion and the anticipated future income when completed. The fact that consent has been gained but not exploited will not however necessarily mean that the proposed development is unviable and that the land value is therefore unaltered. Many landowners will not be willing or particularly able to carry out the development themselves and will prefer to sell the site on to a builder, even though this may mean forfeiting some of the profit. Some owners may couple the sale with a participation in the development gain or offer the land initially on an option, but these variations are unlikely to affect the freehold valuation.

The opportunity to carry out certain developments on farmland is determined not only by the granting of a specific planning consent, but may also be assumed likely according to the zoning of the area in the local Development Plan. Under current procedures, new development is determined essentially by how the site has been allocated in an adopted Plan. The valuer needs to be aware of this even for land lying outside an allocated area, as the Development Plans are revised regularly and there may be a chance of getting the zoning extended at the next review.

Whether a property benefits from a specific consent or from being within a relevant zone, the resultant enhancement in market worth is known as Development Value. The amount of this enhancement depends mostly on the viability of the permitted use but will tend to be slightly less when derived just from the present zoning rather than a specific consent as there is an element of cost, uncertainty and inconvenience in securing an individual consent even within an agreed zone. There are also other cases where value may be influenced by the possibility of a change in use but where it would be misleading to refer to its development value as such. These instances are discussed in Sections 9.1.2 and 9.1.3 below.

9.1.1.1 Valuation of Development Land

In the normal rural context, building development is likely to be on a relatively small scale as in the case of an offlying field or part of a former farmyard. For this it may be sufficient to use comparable market evidence, if there are instances within the locality of land having been sold subject to planning consent for a similar type of development. Just as with other forms of property, allowances will have to be made for any significant variations, including such fundamentals as location and access. For building land there will of course be other factors that could have a direct effect on value. The density of the permitted development, or number of houses per hectare, the anticipated costs of preparing the site and of providing services, the type and likely sale price of the properties when built, are all crucial to the viability of a scheme and therefore to the purchase price of the land itself.

Such matters can be quantified by a calculation known as the 'Residual Method' of valuation. This is effectively a means of assessing the situation from the viewpoint of the developer as potential purchaser. This may be illustrated by the following simple example of a field on which permission has been granted to erect five houses. The first step is to establish, by researching the local market, the current sale price of such houses. From this total is then deducted all the costs that will be incurred in constructing those houses, as shown in the example below.

The figures used in the example are of course simplified for illustrative purposes and will vary greatly according to the individual circumstances. Not only do sale prices of houses vary between different regions and even local sites, but construction costs can also be noticeably higher in some areas, whether because of local pressure and prosperity like in the south-east of England or because of transport and other costs in some of the more remote country districts. Developers' profit, professional fees and marketing costs may also depend on the overall state of the market and local conditions.

	£	£	£
@ £175,000			875,000
Construction costs, including site works:			
5 x 150 m² @ £550 per m²	412,500		
Professional fees @ about 10%			
(architects, planners, quantity surveyors)	41,250		
Total cost of construction		453,750	
Cost of Finance:			
£453,750 ÷ 2 @ 7%		15,881	
Sale costs @ about 4%			
(professional fees, marketing costs)		35,000	
Developer's profit @ £20,000 per house		100,000	
Total cost of development			604,631
Residual sum			270,369
Cost of purchasing land:			
Deduct interest @ 7% for 1 year on £270,369:			
	270,369		
PV of £1 in 1 year @ 7%	0.935		
		252,795	
Net of purchase costs @ 4%			
(stamp duty, agents' and solicitors' fees)		243,072	
Assumed value of site			£243,072

The net amount after deduction of both development and purchase costs is taken to be the price that purchasers would pay for the land with the opportunity to build the houses. The cost of finance is entered here on the basis that the developers will be borrowing funds to meet the progressive cost of the construction and that it would be fair to assume that this would equate to about half of the total cost being borrowed over the whole period of development. The cost of the land has also to be funded and this is therefore netted off the residual sum. The calculation is one to determine what amount must be deducted from the residual sum to allow for the costs of purchase. The table for Present Value of £1 in Parry's (see Section (vi) of Chapter 5) provides the means for working out the cost of finance. The allowance for the assumed purchase expenses of four per cent is made by dividing the net sum by 1.04. For simplicity, the total period of development is taken in this case as being one year. In practice, the final figure of £243,072 might be rounded off to, say, £243,000 in recognition of the various theoretical assumptions that would have been made even in an actual practical case.

9.1.2 Hope Value

The situation of a farm may be such that it is possible that some of the land could be given over to development in the future even though it does not have any current consent for this or lie within the appropriate zone in the Structure Plan. In that case, a prospective purchaser may be encouraged to pay a little over the ordinary agricultural value in the anticipation that he will in time gain an extra return from it. As this is more a matter of speculative judgement rather than an agreed local planning designation, it carries a lesser premium than Development Value and is distinguished often by the separate description of Hope Value.

A property in such a situation may become the subject of an Option Agreement, whereby the landowner is contracted to a development company not to sell some specified land for a particular period. During that time, the company will be expected to pursue the possibility of gaining planning consent for development. If successful, the owner will then be committed to sell the land to the developers under pre-agreed terms, generally on a formula whereby both parties participate in the development gain. Such agreements can be complex and the way in which they might affect the present value of a property will depend on how they are constructed and, of course, on the perceived likelihood of gaining planning permission. The fact that a company has seen fit to take out an option on a site does not guarantee that it will be developed, as it may well be just one of several schemes each competing for consent within that locality. Indeed there are occasions when developers may try to use such agreements to reduce this competition by controlling a number of sites and then pursuing only the one that is the most likely to be granted consent. On the rare occasions when a farm is sold subject to an option agreement, the sale value will depend on whether the vendor intends to retain the proceeds or to assign the contract to the purchaser and also as to how the remaining property may be blighted by the proposed development.

9.1.3 Blight

The development of land will not always result in an enhancement in the original agricultural value. Where such development is carried out on the property under statutory powers, as described in Chapter 13 on

Compulsory Purchase, or on some other property in the immediate locality, the farm may suffer a reduction in value due to the disturbance caused by the new development. An example of this would be the construction of a new road through the land, causing noise and severance, or also the commercial development of an old set of farm buildings which could detract from the residential amenity, and therefore value, of the adjacent farmhouse. Whilst the property may be considered to be blighted by such factors, the expression Planning Blight is also used to describe a situation in which an anticipated development such as a new road causes a drop in value to individual properties likely to be affected by it. Prospective purchasers may become concerned about the effect of blight even before the plans have been agreed and give preference to houses in other locations where such threats do not arise.

To quantify the effects of blight for a valuation is a matter of identifying the impact that the development might have on the property and then interpreting how buyers in the market would currently react to this. With a widespread feature such as a new road, there may be sufficient local examples to give one this market evidence, but in the case of an individual farmyard conversion it will be more a matter of assessing the direct effect that it might have on the enjoyment of the remaining property and the way in which prospective purchasers might respond to that. The same disturbance can have a varying effect on different types of property. For example, increased noise or an interrupted view could have a greater percentage impact on the value of a large period house with grounds and gardens than on a more modestly priced cottage.

Some constraints on value can also arise where properties are situated within certain designated areas. There are, for example, likely to be restrictions upon erecting modern farm buildings within a National Park or an Area of Outstanding Natural Beauty which may in turn have an impact on the full commercial potential of the holding.

9.1.4 Illegal Development

Even after development was regulated under the *Town and Country Planning Act* of 1947, the agricultural sector remained largely exempted, on grounds of its importance to the national economy, and farm buildings could generally be erected without the need for planning consent. More recently, however, agricultural developments have come under closer control and may require a formal consent as laid down in the *General Permitted Development Order 1995*. A valuer needs therefore to establish whether the more modern buildings on a farm required specific consent, if it was obtained and if it is still being used for the kind of purpose defined in the consent. Should there be any infringement, then a prospective purchaser would no doubt adjust his price to allow for the possibility of being enforced to remove or adapt the building and the consequent effect on the farm business.

9.1.5 Agricultural Occupancy

The special concessions for agriculture have extended also to the construction of houses and cottages, in so far as planning permission might be granted for the building of a new house on a farm even though the local policy would not allow any ordinary residential development in such a situation. To gain consent, the applicant would need to establish that there was an actual need for the accommodation, in terms of the size of the agricultural business and the necessity for on-site housing for someone like a cowman who had to live close to the yard. Once built, the property would then always have to be occupied by a person engaged in agriculture. So long as it was still needed for a farm employee, this would have little impact either on the business of the farm or on the valuation. Staffing numbers have however declined over the years and it is now quite usual for such houses or cottages to become surplus to the requirements of the farm. Indeed, this can now affect the principal farmhouse itself as more farms fail to remain viable so that their land has to be sold or let to neighbours leaving the house effectively divorced from the farm business. If these houses or cottages were then to be sold off from the farm, it could only be to people engaged in agriculture. This obviously restricts the market and has to be allowed for in valuation.

There is very little market evidence of sales of such properties, although they are occasionally sold to non-farm workers who somehow may be otherwise engaged in the agricultural sector. Their bid for the house will however be affected by concerns over its ultimate resale and by the attitude of mortgagees. Much will also depend on whether the local planning authority might agree to lift the agricultural restriction, which will depend in turn on the regional circumstances and policies. To allow for all these factors, the valuer can really only make a theoretical discount from the full residential figure of probably between about 25 and 40 per cent. It is always possible, however, that in a real sale situation the price might have to be reduced even further.

9.1.6 Listed Buildings

It is not uncommon for older houses and buildings in the countryside to be listed by the local planning authority as being of special architectural or historic interest. According to the degree of importance and its related grading, such listing will impose upon the owner certain constraints and obligations as regards maintenance or improvements. The effect this may have on value can be both positive and negative. That there can be an enhancement in value, is indicated by the fact that the listing of a property is often used as a selling feature, particularly for fine period houses. Selling agents would hardly promote this feature if it were not considered likely to attract greater interest from potential buyers, despite having to budget for greater expenditure on maintenance and to accept restrictions on modernization and improvements. A listed barn, on the other hand, on an otherwise commercial farm would be recognized as a financial liability as it would have little agricultural use and probably only very limited scope for development, whilst carrying a potential liability for repairs and reinstatement. Other than in exceptional cases where an enforcement notice may have been served requiring the owner to carry out such repairs, the listing itself cannot be directly quantified in valuation terms, and it will be a matter of judgement as to what effect it might have on normal market forces in each individual situation.

9.1.7 Trees and Hedges

The local planning authority has powers also to impose preservation orders on trees where they may be of particular landscape or historic significance. These trees are unlikely to be of such individual worth as to have an impact directly on the value of the farm as a whole, but the preservation orders could well affect the management of certain parts of the land and need therefore to be accounted for in that context.

Hedges too are now protected under *The Hedgerows Regulations* 1997 which can mean that some of these old boundaries cannot be legally removed. This could then prevent the amalgamation of fields that might be considered important for the profitable running of the farm in the future. Whilst again this cannot be quantified precisely, it needs to be recognized as a potential discouragement to potential buyers. This could be particularly so at a time when the market may be heavily influenced by commercial farmers seeking to expand their enterprises and needing to be able to use existing large-scale machinery that would be hampered by small fields.

9.2 Covenants, Rights and Easements

As individual farms cover an extensive area of countryside, often with historical connections, the land may well be subject to particular rights and restrictions. These may be as diverse as a right to stand guns at a shoot on a neighbour's land and a formal easement for a gas or water main. Many of these may be more of an historic or legal nature and of little current consequence, but others can be of greater significance. Their assessment for valuation is based on considerations of cost, inconvenience or risk and the way in which these might affect the attitude of a hypothetical purchaser. To judge this one needs to weigh up the implications of each item, whether to the farm business or to the residential enjoyment of the property or to its long-term freehold security.

For example, a covenant not to create a noise or disturbance to a neighbouring house that might once have been part of the farm could limit the vital use of buildings such as a grain store. Potential buyers may then be inclined to modify their offers to allow for the difficulties and restrictions that this could cause at harvest.

The siting of footpaths can be an example of how the amenity aspect of a farm might be affected as well as the business when, as can often arise, there is one that runs right through the yard and beside the house. Such potential problems or inconvenience require the purchaser to accept a compromise, which may be inevitable for those with limited funds and choice of property. Those with more funds at their disposal need to make fewer compromises and if these stronger buyers are disinclined to pursue a property with this sort of encumbrance, the market price may be weakened accordingly.

A direct calculation of cost might be made when an owner is required to maintain a private water supply to adjoining properties. This can often arise in a country area, particularly where it may have once been part of a traditional estate. Taking on the inescapable responsibility for repairing possibly miles of old iron water pipes may make a potential purchaser work out the cost of replacing it entirely with modern materials. Such cost would then probably have to be deducted from the full market price of the farm and be reflected in the valuation accordingly.

Wayleaves and easements were traditionally agreed on the basis of an annual payment in respect of the ongoing inconvenience caused to the farmer by structures such as pylons. If a property changed hands, the new occupier would become the recipient of these annual payments and thereby be compensated for any annual loss in farm income arising from the structures. Latterly, such payments have often been capitalized so that the original owner gets an initial lump sum and any later occupiers will merely be left with the inconvenience of the structures without compensation. Although not normally amounting to any very significant sums, the existence of a number of such rights across a property and the negative effect on market price should be allowed for in a valuation.

9.3 Small Plots and Paddocks

It is not uncommon that a price needs to be established for a small area of land being offered to an adjacent owner. Examples of this would be where the edge of a field is to be incorporated into a garden or where farmland is going to be converted to paddocks for horses. As the prevailing use of the land was for farming, one might expect that it would be valued as such, by reference to local evidence or to some statistical average figure. In practice, however, the vendor is likely to justify a higher price due to one of two principal reasons.

When a plot of land is being added to a residential property, it may be in order to protect the view from the house, or to enlarge the garden or to provide space for an amenity such as a swimming pool or tennis court. In such cases, the value of the property would tend to be enhanced by the addition, either directly or indirectly because of the protection that it affords against any possible future devaluation from adjacent development. To assess the value of the extra land, one should assess the price that the property would fetch with and without the addition. For example, a fine period house with only a small garden would fail to attract the same calibre of buyer as one that was surrounded by larger grounds and would therefore be worth less. Similarly, an old farmhouse with outbuildings that included some stabling would need also to have some grazing land attached to it in order to be marketed to its full potential. If such extra land is available only separately, its price would be likely to be determined not by what it had been worth to the vendor but by the amount by which it enhances the purchaser's interest.

There will sometimes be evidence of sales of small areas of land taking place, especially of grazing for horses. In that instance, the value would be assessed clearly on the prices being paid for paddocks rather than for ordinary farmland. But, in many other cases, as when adding to garden plots, the transaction may have been privately arranged and not publicized and the valuer will have to rely on using examples of sales of comparable properties that both include and exclude the extra area, and then deduct the assumed value of the one from the other. In other words, market practice indicates that land being sold under such circumstances tends to be priced according to the capital advantage that it would bring to the purchaser rather than any more general considerations. In taking this approach, one must however be able reconcile the reality of the market place with the requirements of the Red Book as regards purchasers with a special interest, as mentioned in Section 1.3. Generally, this would only arise when assessing for OMV the lotted component parts of a property for which there were recognizable opportunities for peripheral sales and when it would therefore be realistic to allow for the potential interest of neighbours.

9.4 Diversification

Changes in the countryside and in the Common Agricultural Policy have encouraged diversification on farms and estates. This may involve the land or the buildings or even the farmhouse itself. In most cases in which part of a property has been converted to a non-agricultural use, the value of that part will be assessed on the viability of the new venture as it stands rather than on any agricultural principles. However, as in the consideration of component parts in Chapter 2.2, this does then have to be looked at within the context of the property as a whole as well. These methods may be explained by reference to the general principles involved and through the illustration of some examples. They are applicable to farm properties that have been partly converted to some alternative use, rather than land that has been given over entirely to a non-agricultural use. In this latter case, the property would be valued by reference to the market for that kind of enterprise and to its commercial viability. Where the development still forms part of a farm, these same considerations will apply but will have to be balanced against the impact that it might have on the value of the property as a farm.

9.4.1 Regulations

Most farm diversification schemes will have been implemented by the farmer himself using his own labour and trusting perhaps to his own expertise. This is partly by the nature of such businesses and partly out of necessity from limited capital resources. It suggests that particular care may need to be shown by the valuer in checking that situation is not prejudiced by a lack of the necessary planning consents and of other by-law approvals such as building regulations and health and safety requirements. Any shortcomings in this respect would affect the value by either the cost of remedying the work or, more seriously, by the risk of being required to restore the premises to their original state.

9.4.2 Viability

Diversification in the form of a commercial venture must be capable of showing a profitable return if it is to have a positive value. Some schemes may possibly be only for amenity or as a hobby, in which case their intrinsic value would be determined by whether or not they enhance the value of the whole. For a commercial diversification to be *capable* of showing a profitable return need not mean that it is actually doing so under the present management, but that a prospective purchaser could be expected to run it in a viable manner. It can be difficult for someone whose training and experience has been in farming rather than in the field of the newly introduced business to manage it as successfully as someone who is more familiar with that particular sector. The fact therefore that a farm includes an unprofitable venture need not automatically be discounted in value as it could still attract buyers with the necessary skills to make it show a positive return.

An example of this might be when a farmer has gained planning consent to convert part of the yard into a garden centre, but has failed to secure any trading profit from it. If the whole property were offered for sale, that particular part could be of great interest to horticultural companies who had been unable to find a suitable site in the area on which to develop their own centre. This presupposes that, as such companies are unlikely to buy complete farms just to gain a retail outlet, the whole property could be lotted so as to separate out the garden centre and leave the agricultural part still as a sufficiently attractive and viable entity – or otherwise as a combination of saleable lots.

The valuer needs however to consider the question of viability on a much wider perspective than when dealing with an ordinary farm. In the agricultural context, which was discussed in Chapter 2, these questions focused on the outlook for farming generally and then on the quality, size and structure of the farm itself, without reference necessarily to what other farms in the neighbourhood were like or were doing. When assessing a property on which some alternative business has been introduced, it does however become relevant to look further afield. If, for example, some of the buildings have been converted to a farm shop and are currently attracting good business, this could be threatened if another farmer within the area opened a rival shop and had the advantage of a more accessible location, better buildings and more resources. The original farmer would lose custom to the better-placed shop and the longer-term viability of the enterprise would be compromised. It is of course upon such longer-term considerations that valuations depend. Whilst one cannot foresee whether anyone might start up a rival venture, the valuation may be influenced by whether the site has disadvantages and by any other limitations of the buildings or the location. If the property can be identified as being well suited to its converted use, it will be less vulnerable to potential competition and would therefore attract a more confident attitude among prospective buyers.

Unlike farming which is dependent in the long run on international factors such as the Common Agricultural Policy and World Trade Agreements and in the shorter term on the weather and world prices, many rural business ventures can be affected by domestic economic conditions and on medium-term weather patterns. On-farm enterprises are invariably in relatively remote rural areas that rely upon private cars, sometimes travelling great distances from the more populous towns. Those that depend on a retail trade or on visitors in the form of tourists or overnight guests can be directly affected by a rise in petrol prices or by a growing trend to holidaying abroad. When the new venture represents a substantial part of the property and of the total farm business, the investment value could be influenced by potential changes in these basic national patterns.

These brief examples show that a valuation of a farm on which there has been some non-agricultural diversification may need to incorporate an approach that one would more normally associate with a management consultancy report.

Rural diversification is receiving a growing amount of political and financial encouragement under the European Development Programmes as well through national measures. This engenders a growing expectation that farm properties should offer opportunities for development and diversification. While these expectations may not always be fulfilled in practice, they do need to be identified in a valuation appraisal. If potential purchasers are likely to be influenced by a perceived possibility of some alternative enterprise on the farm, then the effect of that interest on price will have to be accounted for within the value.

9.4.3 Agricultural Diversification

There will be farms where new enterprises have been introduced which are still a form of agricultural production, albeit different to the previous core businesses and involving new investment in plant or buildings as well as stock. Examples of this would be conversion to organic production, the introduction of exotic livestock such as llamas or new crops like soft fruits for a 'pick your own' operation. The same business principles described above will still apply when making a valuation appraisal. Although problems over planning and local competition may not be so crucial, the questions of expertise and market security can be just as relevant, particularly in the eyes of the prospective purchaser who may not expect to have the necessary expertise or confidence to continue with an unusual venture that requires a substantial capital commitment.

9.4.4 Valuation Approach

Diversification in a rural context will in most cases be as a relatively minor part of a larger farming enterprise. The initial valuation approach will then be to consider the whole from the viewpoint of a potential purchaser. Whether the opportunity for running an alternative business alongside the farm will attract an enhanced bid will depend on its current viability and on the degree of specialization and management time that it may require. If it is seen as a profitable and attractive venture, then the general market attitude is likely to add a value to it just as if it were a component part as discussed in Section 2.2. Where the enterprise is of a larger scale and more independent of the rest of the farm, the income that it generates may be assessed as a means of increasing the profitability and borrowing power and therefore also the price of the property as a whole. If the business is let or franchised for a period of years, it would be appropriate to capitalize the rental income and add the resulting sum to the value of the whole, using the principles of the investment market as described in Chapter 5 and as illustrated below.

Consider the example of an arable/stock farm of 300 hectares, with a six-bedroomed farmhouse, a pair of cottages and a modern range of buildings. The basic valuation might be made up as follows:

Farmhouse	£350,000
Cottages @ £125,000	£250,000
200 hectares of arable land @ £4,900	£980,000
100 hectares of grass land @ £4,200	£420,000
Total agricultural and residential value	£2,000,000

(The land prices have in this instance been taken as inclusive of the fact that the farm is well equipped with buildings, although these buildings could also have been quantified separately as in the example in Figure 2.1 in Chapter 2).

If the property were also to include a traditional barn that had been converted to offices and was now let under a commercial lease at a current rack rent of £15,000 per annum, this might be incorporated in the valuation on an investment basis:

	Agricultural and residential value		£2,000,000
	Office rent	£ 15,000	
less	Maintenance costs, say	£ 3,000	
	Net rental income	£ 12,000	
	YP @ 10% in perpetuity	10	
	Investment value of barn conversion		£ 120,000
	Total		£2,120,000

The YP of 10 would have been selected as a proper reflection of market attitudes to a small commercial letting of this kind, bearing in mind all the usual factors such as the standard of accommodation, location, current rent and terms for review, and tenants' covenant. The earlier figure of £2,000,000 for the agricultural and residential elements has, in this instance, been carried over in full, although it could be that the existence of a commercial business in the farmyard might spoil the amenity of the farm as a whole in which case the value attributed to the house may have to be revised accordingly.

Although these conventional valuation methods can be applied to the non-agricultural business elements on a farm or estate, one must be aware that there is far less activity or evidence of this kind in the rural market place than in the commercial sector. The numerical outcome of an investment calculation should therefore always be reviewed against the context of the factors described in Sections 9.4 to 9.4.3 above.

As a further illustration, one might now assume that the property benefits from another source of non-farming income in the form of bed and breakfast accommodation and that this also grosses £15,000 per annum. It would be possible of course to treat this in the same way as the office rent and apply an appropriate rate of capitalization. In practice, however, one would be likely to take a more general approach and try to establish a figure by which this facility might enhance the overall market price. Unlike a formal commercial lease, which produces a relatively assured investment income, the bed and breakfast business depends on the occupier's own efforts and abilities. It also has a direct impact on the enjoyment of the farmhouse as a family home. It could be therefore that whilst one type of buyer might see it as a means of earning the equivalent of a further £50 per hectare from the farm, another would consider it as reducing the attraction of the property as a residential farm. The former might use the bed and breakfast opportunity as a means to justify a higher bid, whereas the latter would be inclined to discount it entirely. In either case, this feature would probably be best incorporated within the figure for the farmhouse, which would then reflect either its full residential appeal or the added commercial potential from being able to offer bed and breakfast accommodation. The final outcome will be determined largely by the nature of the property and of the market place within it would be likely to be sold.

9.4.4.1 Financial Approach

In cases where the diversified business represents a significant investment, it may be appropriate to assess that part of the property in terms of its financial implications. One such basis is the Net Current Replacement Cost (NCRC) which effectively defines the price that would have to be paid in the open market should one need to replace all the facilities within which the business is housed. This uses the same approach as for Depreciation Replacement Cost, (DRC) in Section 2.5, except that there is no discount for depreciation or obsolescence; the distinction being that the Current Cost method is applied to situations where the age and design of the buildings are of lesser consequence in market terms.

This might arise in the case of traditional barns being used as workshops by craftsmen or a farm shop, as opposed to specialist livestock units. In that latter market there is likely to be a greater price differential between new or old buildings as buyers can be directly influenced by the age and design of the premises. In the former instance, however, these matters are of less importance and price will be determined more by other market issues such as location, size and suitability, although allowance would of course still need to be made for the actual condition of the property. To determine this theoretical cost of replacement will therefore depend both on the evidence of sales of comparable properties and also on market attitudes to that type of premises. Where there is an identifiable premium for more modern buildings, the DRC may be the more appropriate, but otherwise a straight NCRC might be used.

Another similar approach is that of Existing Use Value under which the property is assessed as for OMV but with the additional proviso that it can only be used in the foreseeable future for its present purposes. This can be particularly relevant in situations where the diversification has been permitted under strict planning controls, so that it would be unlikely that this could be readily changed to a different usage.

Check-list – Development and Diversification:

- Check whether planning consents exist but have not been implemented; whether relevant developments have the necessary consents; whether any new development is likely to be permitted on the property, or elsewhere so as to disadvantage the property.
- Distinguish between Development and Hope Value, and Blight.
- Agricultural restrictions and Listing.
- Diversification, and whether this always adds value.

CHAPTER 10

Woodland

In the context of agricultural property, woods will normally form an integral part of a farm and estate and be valued as such, as mentioned in Section 2.3.4. Where the land includes areas of commercial forestry its appraisal will involve a specialized knowledge of the market and of silviculture which may well be beyond the experience of most rural practitioners. It is, however, useful to know the basis upon which such valuations can be made.

The concept of commercial woodlands may conjure up images of large areas of regimented conifers spread across the remoter upland areas of the country, but there is also an active market in small blocks of mixed woods. The valuation of these smaller parcels is made on the same basis as that of an ordinary farm, namely assessing market comparables and making the necessary adjustments according to the individual circumstances. This method can also be used for some of the larger forestry properties, although in those cases other techniques are likely to be employed as well.

10.1 Market Evidence

In the farmland market, whilst often there may be a shortage of market evidence, it does at least get publicized in the main farming papers, as mentioned in Chapter 3. The forestry market is more specialized and gets only occasional mention in the agricultural press so the valuer will therefore need to look further for the necessary information. Woodland sales tend to be handled by a small number of specialist agents who issue bulletins occasionally of the selection of properties that they have on offer, and from which one may glean current asking prices for the different types and sizes of investments in various locations.

There is also an index of forestry investment produced annually by The Investment Property Databank (IPD) in conjunction with FDPSavills and Fountains and a number of other interested parties. This gives a comprehensive analysis of investment returns, rather than prices, but does none the less provide additional information on market trends.

10.1.1 Comparables

Many of the small woods that are bought and sold as investments will have little real commercial potential and will be held for reasons of amenity or long-term appreciation or ultimate development. The market price is therefore more a matter of judging potential demand for the property as a whole than of calculating future returns from timber production and this will be determined by the evidence of recent similar transactions. In making comparisons, however, one must be aware of what type and age of trees are growing on the site and whether these might offer a commercial or sporting potential that another property with an apparently similar location and aspect would not have.

10.2 Types of Woodland

In market terms there are two essential categories of trees; broadleaved or conifer. A property may comprise just one of these or a mix of the two. Within these there are of course many different types and species, but there can be distinct market characteristics according just to the basic type. Broadleaves take longer to mature and therefore require a longer-term investor and tend to be in lowland areas with relatively higher site values. Conifers, on the other hand grow more quickly and so produce an earlier return and can thrive on low-cost upland areas. Broadleaves attract a higher rate of grant, and the commercial varieties such as oak, beech and ash produce a more valuable timber than most conifers.

As even conifers take around 70 years to mature and will be unlikely to produce any income other than grant for the first 20 years or so, the age of the trees is a significant factor in market value. The larger plantations may comprise a mix of ages allowing a fairly even cash flow as individual compartments become due for thinning or felling at different times, but in most cases there will be a time factor that influences value.

10.3 Grants and Woodland Policy

An extensive area of Britain's forests is in public ownership under the auspices of the Forestry Commission. In order to encourage tree planting in the private sector, a range of grants is available with the purpose of reducing the financial burden that owners of new plantations would otherwise have to bear until their crops reached maturity. Most of these grants are now paid directly for planting work, either as a single cash amount or in two stages over five years. Although vital to the viability of establishing new woods or to replanting areas that have been clear felled, they do not normally have a direct influence on the valuation of existing woodlands. Some grants are paid annually for specific management work such as maintaining environmental features.

Although these are relatively limited both in their application and in the amounts offered, they should be noted in a valuation because penalties might be incurred if the management agreements were broken at some future stage. This can be significant in the case where farmland has been planted with trees and where a more substantial annual payment is being made for up to 15 years, according to circumstances.

This last scenario would be for agreements made under the Farm *Woodland Premium Scheme*, whereas existing woodlands come under the *Woodland Grant Schemes*. Details of rates of grant and the conditions applied to them can be obtained from the Forestry Authority at one of their regional offices or the headquarters which are in Edinburgh.

While grants and certain tax concessions are offered as an encouragement to plant trees, there is also an element of enforcement on owners of existing woodland. No trees may be felled without permission from the Forestry Authority. This applies to any timber amounting to more than the equivalent of two large trees (or five cubic metres) in any three-month period, so that for even the smallest commercial operation such consent will be required. When there is a valid reason for the trees to be removed, permission will be granted in the form of a felling licence. This will, however, normally be coupled with a requirement that the land be replanted with an agreed type of tree or mix of trees. When valuing woodland, therefore, it will usually be necessary to assume that the property will remain as such in perpetuity and that there is no likelihood that it might be converted to farmland after the mature crop has been felled. The possibility of gaining planning consent for development then this would override the normal constraints of a felling licence and the valuer would then follow the procedures discussed in Chapter 9.

The implications of tax on the woodland market are dealt with in Section 10.5.3.2 below, but in the present context it may be noted that no income tax is payable on the proceeds of sales of timber provided the woods are run as a business; that growing timber is exempted from Capital Gains Tax; and that Business Property Relief is available against Inheritance Tax.

10.4 Yield Classes

Although essentially slow growing, trees do mature more quickly under certain conditions, according to a combination of climate and location for each species. This is identified by a Yield Class which is a measure effectively of the rate at which a certain type of tree can be expected to grow on a particular site. A higher Yield Class implies a shorter rotation, or time to maturity, and potentially therefore a higher site value.

10.5 Methods of Valuation

10.5.1 Using Market Comparables

In certain cases it may be possible to compare the woodland being valued to other properties that have recently been sold or are currently being offered for sale at specific asking prices. This can certainly apply to the smaller type of investment wood referred to earlier, where the main market features may be ones of size and location rather than the age and detail of the timber being grown there. It can also be used with some of the larger commercial plantations, although it may be more difficult to find the appropriate market evidence and there are then likely to be a greater number of technical variables to consider too. It will then often be necessary also to assess the individual woodland in more detail as follows.

10.5.2 Land and Timber

One can begin by separating out the site, or 'prairie', value from the standing timber. The land will be assessed as planting land for which there should normally be sufficient market evidence, although certain assumptions may have to be made particularly about the infrastructure regarding, for example, roads and drainage.

The method for valuing the growing timber will then depend on the age of the trees at the time. When they are nearing maturity, the volume of marketable timber is measured and priced according to what it could fetch from merchants as a standing crop. This figure is then added to the assumed value of the site as planting land to give an approximation of market value for the whole property.

If the trees are still young, the valuation is based on a calculation of the expenditure that will have been incurred in establishing the plantation. This will be an accumulation of the cost of preparing the ground, buying and planting the original seedlings and maintaining and protecting them meanwhile. This total is again added to the estimated value of the site as planting land.

Wherever possible these figures should be then checked against such evidence as there may be of actual transactions of woodland properties of that kind. There may, however, often be occasions when this method of assessment will have to suffice as an acceptable basis of valuation.

When the trees are semi-mature, the choice of which method of valuation to use becomes a matter of judgement. It is, however, most likely to involve an approximation of the first calculation using estimates of what the final volume might be and discounting it over the period of intervening years. This estimate can be made with some degree of accuracy by referring to the Yield Classes mentioned in Section 10.4 above. If one knows the year when the trees were planted and can establish by means of the Yield Class and special Management Tables the rate at which they will reach maturity, then one can calculate both the final volume and the time which needs to be discounted until that stage has been reached. The rate of discount is also a matter of judgement but will probably be taken to approximate to the level of Internal Rate of Return being sought by investors at the time, as referred to in Section 10.5.3 below.

These calculations are based on the volume and value of the final crop of felled timber, but there may be circumstances where there is a sufficient worth in the interim thinnings for these to be included as well. One can deduce the appropriate figures for this from the Management Tables and then incorporate the additional figures into the basic valuation where it is thought that the nature of the crop is such as to produce a significant income from thinnings.

10.5.2.1 Measurement of Volume of Timber

Reference has been made above to measuring the volume of standing timber. This is a exercise that requires a degree of knowledge and practice and is more fully explained in Forestry textbooks, notably *Practical Forestry for the Agent and Surveyor* by Cyril Hart and *Forest Mensuration* published by The Stationery Office for the Forestry Commission. For property valuations involving even-aged groups or 'stands' of commercially planted trees, it is customary to measure the trees in a sample plot and to multiply this out for the whole area. The heights of marketable timber of the trees within the plot are estimated and their diameters measured at 'breast height' and volumes then calculated by reference to special tables. In a general land management context, the plots will be set out and measured by tape but for commercial forestry a more sophisticated process is often used whereby the trees are assessed visually by means of an optical device called a relascope.

10.5.2.2 Establishing the Age of a Plantation.

The age of the plantation, which is needed when working with Yield Classes as mentioned above, can normally be established from the landowner's records which will have been required to be maintained in conjunction with the grant schemes. If such records, or compartment notes as they are known, are not available and it is important to have the exact age of a seemingly even 'stand' of trees, one may resort to finding this out by use of a chain saw! By felling one tree and counting the rings in the exposed cut, one will arrive at the precise number of years that it has been growing.

The cost of establishment used for the valuation of younger plantations will need to be net of grants that are receivable for planting and maintaining such woods. In assessing newly established woods that qualify for annual management grants, such as those planted on farmland or with special environmental potential, one can consider capitalizing that grant income in lieu of any actual market evidence.

10.5.3 Commercial Considerations

To determine the value of a forestry property, one follows the same principles as for any other form of land, although it can be more difficult for the non-specialist to identify the physical features and the market factors. Unlike agricultural property, however, there is a special section within the Red Book dealing with Forestry and Woodland Properties (Guidance Note 12).

10.5.3.1 Property Features

The assessment of the calibre and potential of the trees themselves has been described briefly above, but in conjunction with this the valuer will also need to consider the way in which the property features will affect the value as a whole. Some of these will be similar to those for farms and estates but often with a different emphasis and impact.

For example, certain locations are more favoured than others and one of the factors by which this will be determined is the distance of the property from the nearest processing or shipping facilities, bearing in mind that the transport of timber is a heavy cost. Also, although most commercial woods are in upland areas, the altitude and exposure of the site can be crucial not only in terms of hampering the growth of the young plants but also because of the danger of maturing trees being devastated by wind. In fact this is of such significance that a special measure has been developed by the Forestry Commission in the form of Windthrow Hazard Classification. The risk from windthrow can be mitigated by choice of species and careful management, but a high classification would none the less have a negative bearing on value.

Access is another feature that can be more significant in forestry than to farmland. First, commercial woods are often in remote situations and yet require hard tracks for the movement of heavy equipment and of timber. If these have to be built from new, it would be at a significant cost which might be discounted in the market value. Second, public access is an increasing issue, arising from a number of factors including the policy of selling Forestry Commission woods. Such access can restrict shooting in the woods and increase the chances of fire damage. A forest that is accessible by its proximity to towns, motorways or to established holiday areas, may therefore again be discounted to a degree in the market.

Wayleaves, as a further example, are usually only a minor inconvenience on farmland, but can create swathes of sterile areas in woods as no trees can be planted along the routes of gas mains or electricity lines. Fencing liabilities and drainage are also both matters that come up for consideration on farms but which can again be of greater significance when put into the context of a large expanse of wet upland forestry.

It is not feasible in this context to give an exhaustive list of all such features that might have a special bearing on forestry values, but the above examples should serve to emphasize the different areas of technical awareness that need to be considered in conjunction with the more general property appraisal.

10.5.3.2 Market Factors

Due to the fact that there is so little commentary published about the sale of forestry property, it can be difficult to gauge market attitudes unless one is actively engaged within the sector.

The forestry sector did previously benefit from significant tax advantages particularly in the form of being able to offset costs against profits from elsewhere. This in turn attracted investors who effectively dominated the market. These concessions were rescinded under the *Finance Act* 1988 and phased out by April 1993. There has since then been a decline in activity as private purchasers focus on the smaller amenity type of wood and as the sale of larger commercial areas becomes more dependent on institutional and other funds. To value the larger blocks of woodland one needs therefore to be able to define the potential market in much the same as for agricultural investment land, as described in Chapter 4. Due to its specialized nature, this is a matter on which most rural practitioners would expect to have only an outline knowledge and in this context it is sufficient to highlight just the main influences.

In general, the annual returns from forestry depend to some extent upon the level of sterling and on the economy as a whole. For many years now, timber consumption in Britain has been met largely by imports (amounting to well over 80 per cent of total demand) so that prices of home grown wood are closely linked to the equivalent from abroad. As sterling rises, domestic timber prices fall and with them some of the demand for investments. When the UK economy is buoyant, there is greater demand for construction timber and the interest in forestry investment may rise accordingly. These reactions occur despite the long-term nature of such an investment. On the other hand, recent trends have meant that investors have been prepared to pay relatively higher prices for commercial woods that are closer to maturity than for younger plantations that could be bought at lower prices and which would therefore show a better overall return if held over the full rotation.

Market prices depend also on the balance between supply and demand. In the 1990s, following the impact of the tax changes mentioned above, this has been relatively stable but it is possible that new initiatives may arise for the sale of more Forestry Commission holdings. The market would of course be affected by such disposals and valuers will need to be aware of the prevailing situation on this.

10.5.3.3 Yields and Prices

In the farmland market, vacant land tends to be referred to in terms of a price per hectare or acre whereas tenanted investment farms are often quoted instead as reflecting a particular initial yield. The same approach applies to the forestry market, although with different distinctions.

Smaller woods, that are of interest mostly to private investors, are generally assessed in terms of a price per hectare. Larger plantations with a more commercial content are, however, considered on the basis of what the future income stream should represent by way of a return on that particular calibre of investment, which in turn determines its price. This end of the market deals therefore more in terms of Internal Rates of Return (IRR) based upon a Discounted Cash Flow analysis, as described for investment properties in Section (xii) of Chapter 5. As the term for which the valuation is being carried out can stretch over several decades in the case of a mid-age plantation, the rate per cent used for discounting can be of crucial significance to the current capital figure being derived. It is therefore particularly important to work with the most appropriate figure for IRR, which in turn requires a full awareness of the prevailing trends in the forestry investment market.

10.5.4 Insurance

The valuation of woods for insurance will follow the same principles as those just described, although excluding the value of the site itself. It can be argued in this context that when assessing the cost of establishment of a young plantation, one should include an item for the delay that the owner will incur by having to replant the crop. For example, a plantation of 10 year old conifers may be assumed to reach maturity in 60 years' time. If it is devastated by fire and has to be replanted, the owner will have to accept a delay of 10 years until the new trees achieve their natural maturity of 70 years. The loss in investment return represented by this postponement may need to be discounted by the appropriate financial calculation.

Due to the relative complexity of valuing woodlands, some insurance companies work with their own standard set of figures for specified age-groups of different types of trees in both lowland or upland areas. These can be a convenient check against market values, although tend to be a little higher than the general market.

Check-list – Woodland:

- Principles of commercial forestry valuation; limited market evidence.
- Separate assessment of land and trees, and different approach according to age.
- Grants and tax advantages
- Methods of measuring performance and quality.
- Private investments and commercial plantations – price per hectare and IRR.
- Principles of insurance valuations.

CHAPTER 11

Sporting Rights

For valuation purposes sporting rights are treated in some ways in a similar manner to woods, in that on most farms and estates they will be an ancillary feature that serves to enhance the overall value of the property, as mentioned in Section 2.3.4, although there will also be properties where shooting and fishing are a major part of the economic worth and will have to be assessed separately. In either case, the valuer will be concerned with those traditional sporting rights that can be owned by a landowner and exclude others such as hunting which do not normally attract any commercial payment.

In lowland areas, shooting will concentrate on pheasant and partridge, with some roe deer stalking and also wildfowling, and in some upland districts such as Scotland and north England, this will extend to grouse shooting and red deer stalking. Fishing divides into two categories, game and coarse, with the former being either for salmon or trout.

When the sporting rights can be run on a commercial scale, their value will be assessed partly upon the range and quality of essential features such as terrain or woodland cover and also upon the records of what numbers of game have been 'bagged' in the past. In some cases, the shooting and fishing may have been enjoyed by the owner without any financial return, but there will also be market evidence of these facilities being let out either on a daily basis or on a lease for one or more seasons. For most low ground sport, as described below, the market value will be derived from a capitalization of the potential income from such lettings following the principles of investment calculations as explained in Chapter 5. It is, however, difficult to know precisely what Years' Purchase to apply as there is usually insufficient actual market evidence to guide one on this and much will depend on the circumstances of the property and the quality of the sporting. On the whole, YPs of between about five and ten have tended to be used for that purpose. This will give a capital sum that can be added on to the basic value of the land, as a component part using the method described in Section 2.2. There will be occasions, however, when the fact that a property has sporting facilities may enhance the value of the whole by more than the amount attributed to those facilities alone. Such enhancement is to be expected on residential estates in attractive country areas accessible to major towns. In capitalizing the potential rent, allowance will need to be made for the cost of keepering and stocking and whether this is the responsibility of the landlord or of the tenant. In some cases this may even include the provision of a cottage for the gamekeeper.

Lowland shooting on a commercial basis can involve intensive management, with possible adverse affects on the surrounding farmland, whether in-hand or let. This can arise from having to allow access over tenants' land, from requiring corners of arable ground to be sown with 'game crops' to shelter and feed the birds, and from the damage done to the farm crops by the pheasants themselves. Such issues may have more of an inconvenience rather than a financial loss, but there are occasions when they have to be taken into account in assessing the overall value of a shoot.

In the north of Britain, where sporting can be the major purpose of an estate, the value of the entire property will be assessed according to capital sums attributed to the game itself, as explained in Section 11.1.2 below. These amounts depend on a diversity of factors, including the location, the quality of the sport on that particular property and whether it has been plentiful or reduced in the preceding years and on the economy as a whole in so far as it affects the spending power of potential tenants. In contrast to most pheasant shooting or trout fishing, highland facilities such as grouse shooting, deer stalking or salmon fishing may be let on a weekly basis and combined with accommodation in a 'Lodge' with supporting staff, both domestic and sporting.

The general principles can be illustrated by looking at examples of the main enterprises, as follows.

11.1 Shooting

11.1.1 Pheasant and Partridge

The value of shooting depends upon the scale of the property, its location and layout, and on the past management and records. For a shoot to have a commercial value, the land needs to be extensive enough to provide scope for several days' shooting for probably not less than about eight guns. This means that it is unlikely to be of direct relevance to farms of less than about 300 hectares, although an opportunity for occasional shooting on a non-commercial basis can still add to the overall value of smaller properties as part of their general amenity, as discussed in Chapter 2.3.4. Smaller farms may well also be linked with neighbouring land to form a larger viable shoot, but as this will probably be on a short-term renewable basis it would not be appropriate to include it in a freehold valuation. On estates and larger farms, it will be a matter of identifying whether the terrain is particularly suited to pheasant or partridge. This requires some knowledge of shooting, but in summary will mean woodland and scrub cover for pheasant, reasonably spread out and with some shape to the landscape and open, dry arable ground for partridge. Apart from the physical features of the property, one can also gain further indications of the calibre of the shoot by looking at its recent history. This will include a record of the number of birds 'put down', or introduced to the property, the number of days on which shooting took place, the number of guns (and whether double or single), the number of drives and the totals shot per day.

If there are no formal records of the recent management of a shoot and of the 'bag' or number of birds shot, one will need to identify whether the property has the characteristics to attract keen purchasers or tenants. This will be a matter of being able to recognize the general factors described in the previous paragraph and to assess the more technical potential. For pheasant shooting, this includes the layout and direction of the 'drives' or areas of cover interspersed with clearer ground and the height at which the birds overfly the guns, or are 'presented' to them. The contours of the land can be of importance too in that where, for example, the woodland cover overlooks a small valley then the guns when standing in the lower ground are faced with higher and therefore more challenging birds. A further consideration is then whether such cover is spaced across the land in such a way that birds can be driven over the same stand of guns twice from opposite directions.

Partridge shooting is less concerned with woodland cover and takes place on more open ground which is the more natural habitat for these birds. Such country will not necessarily offer opportunity for high birds, as in the case of the wooded valley, but nor is this required for partridge which tend to fly lower over the ground anyway but are faster than pheasant.

Location is also important, namely as to whether the property is sufficiently accessible. A shoot that has all the best ingredients but which is too remote to be visited easily by city dwellers may be worth less in both rent and capital terms than a less well suited property that is within easier range of urban areas. Where pheasant and partridge shooting is not retained for the exclusive enjoyment of the landowner, it may be let out on a daily basis or on longer-term agreements. The number of days let, and the number of guns accommodated, will depend on the scale and layout of the property. Rents for short-term lets are assessed by the day or in terms of numbers shot, and tend to be for syndicates bringing the full number of allocated guns, rather than individuals.

11.1.2 Grouse

Grouse are confined to certain types of moorland in parts of Scotland and northern England and also, to a much lesser degree, to a few areas of mid-Wales. The value of such a moor will therefore depend largely upon how well suited it is as a habitat for the grouse. This will be largely a matter of natural topography but may also have involved active management in such matters as the control of heather and of sheep grazing and of natural predators. Grouse cannot be reared in the way that pheasants may be on a lowland shoot, and the numbers recorded will therefore be just of those that have been shot each year.

As the moor tends to have very little alternative commercial value, probably only as sheep grazing, its freehold value is often calculated by reference solely to the number of grouse. The market evidence tends therefore to assess prices paid for grouse shooting properties in terms of a capital value on each pair or 'brace' of birds shot over an average of previous years. As there are only few actual examples of such moors being bought or sold at any one time, these figures are reckoned at effectively a standard rate for each region and varied then when appropriate according to the particular features of an individual property. Apart from topography, there will be questions of size and location and also as to whether the moor is 'driven' or 'walked-up'. In the former case, the shoot has been organized so that the guns are positioned in specially constructed 'butts' or hides and the birds are driven towards them by beaters. This facility can command a premium over a shoot where the guns will need to walk the moor for themselves. It is a convention that grouse are counted for the records, and therefore for valuation, in pairs or as a 'brace' rather than singly. Being an essentially wild bird and therefore not reared and fed in the way that pheasants can be, grouse are susceptible to disease and adverse weather, and records of bags can be affected accordingly. Whilst any undue variations may be balanced out by using an average over a number of years, the valuer must be able to account for temporary declines and for the effect this may have on the level of demand from sportsmen.

Again, as for pheasant and partridge shooting, much will depend on the terrain and its natural suitability for such birds. However, in the case of grouse, this will rely additionally on whether the moor has been properly managed, with in particular the control of heather to allow for successful breeding of the birds. Rents for moors that are let are generally on the basis of the number of brace shot, the rate varying of course according to the quality of the terrain and the past records.

It remains still to be seen whether the value of grouse moors in more populous areas may be affected by the introduction of the *Countryside and Rights of Way Act 2000* which is to allow unrestricted access to moorland.

11.1.3 Red Deer Stalking

The red deer is indigenous to the highlands of Scotland in areas described in property terms as deer forests – although they tend to be bare hillsides without the trees that the name might imply. As in the case of grouse, this terrain has little alternative commercial value and is therefore valued by numbers of stags shot in the past and depends upon some of the same general factors of topography, location and scale as well as records of previous bags. Stags are shot by the one method of stalking and depend more upon the ability of the local game keeper or 'stalker' than upon any distinctive management of the property. Stalking is let by the day and priced in terms of the value of a stag.

11.1.4 Lowland Deer

Deer stalking in lowland woods has a limited commercial appeal and would not normally be assessed as a separate calculation. In most cases, therefore, the presence of roe, fallow or muntjac deer and a suitable location on which to shoot them would be noted as part of the overall attraction of the property rather than as a value in itself. There are properties that have lent themselves well to roe deer stalking and which appeal to visitors from continental Europe and where worthwhile lettings can be made, usually on a weekly basis. Rents are likely to be based on a price per buck shot.

11.1.5 Wildfowling

Wildfowling, or the shooting of duck and geese, takes place either on coastal areas of mudflats and estuaries or at 'flightponds' which are lakes or ponds situated inland. The latter may often be seen as an additional feature to enhance the overall value of an estate shoot and records of bags are likely to be kept so that they can be included in the assessment of the annual rent that might be charged, and therefore capitalized. With the coastal areas, there is the distinction that land may have little commercial use other than shooting, and also that the sporting rights can be separate from the ownership of the land itself.

Shooting records may also include mention of small numbers of woodcock and snipe, particularly where the land is near to water. Although these do not normally have a significant value in themselves they can serve to enhance the assessment of the whole.

11.2 Fishing

11.2.1 Salmon

The salmon is a migratory fish that uses particular rivers in Britain for breeding. These are necessarily of a certain size and found mostly in Scotland but also in parts of Wales and the south-west of England. As with some of the shooting game mentioned above, salmon fishing is assessed in terms of the capital value of the average number of fish caught in previous years. The final figure is then influenced by the particular advantages of the river in question, including location and ease of access, as well as also an element of reputation or fashion. Values tend to be greater on the well-known rivers of Scotland than on those where the fish may be fewer in number and smaller in size. There is little physical management of such rivers and therefore little value added by this, other than the provision of a 'lodge' and possibly also the calibre of the keeper or 'ghillie' who assists in the actual fishing itself.

11.2.2 Sea Trout

Rights to fish sea trout, also known as salmon trout, are assessed along similar principles to salmon. Confined mostly to the lower reaches of tidal rivers, sea trout represent a smaller and less prestigious market.

11.2.3 Trout

Whilst the brown trout was also originally an indigenous fish, it has over the years been augmented through the introduction of outside stock in the form of rainbow trout. Both are only able to thrive in particular types of water, associated with chalkland, and whilst the more sought-after fishing is on streams and rivers, trout can also be found on lakes and lochs. Trout fishing rights tend to be assessed in terms of the length of river over which they pertain, rather than the numbers caught, and whether these rights apply to the whole of that stretch or just one side of it, described as double or single bank. Once again, location, reputation and fashion are determinants of value but in this case these values may be enhanced also by good management of the river, whether in terms of stocking or of maintenance of weirs or of weed clearing. The valuation of a lake used for trout fishing will tend to be on the basis of the commercial return obtained from it and associated business factors such as location and the quality of the facilties.

11.2.4 Coarse Fishing

Coarse fish are available in a much wider range of rivers, canals and lakes than game fish, but attract a correspondingly wide following. In a farm or estate context, however, this will not normally represent a significant value due to the relatively limited rental income derived from angling. It would take note of the facility offered and the income that it produces rather than the particular worth of the stretch of water when compared to any other. Location will none the less be important, particular if it is near an area where angling is popular. There may also be some influence from the fact that it has been well managed.

11.3 Hunting

As both fox and stag hunting have not traditionally attracted any commercial return to landowners, they would not be reflected within a valuation of a farm or estate over which these sports may have taken place. To that extent, the likely ban on this sport would not be expected to have any direct influence on the values of properties in traditional hunting country. In a more general market context, however, there are areas where landed residential properties may have commanded a premium on the strength of their situation within a renowned hunting district and this would of course no longer apply if the sport were abolished.

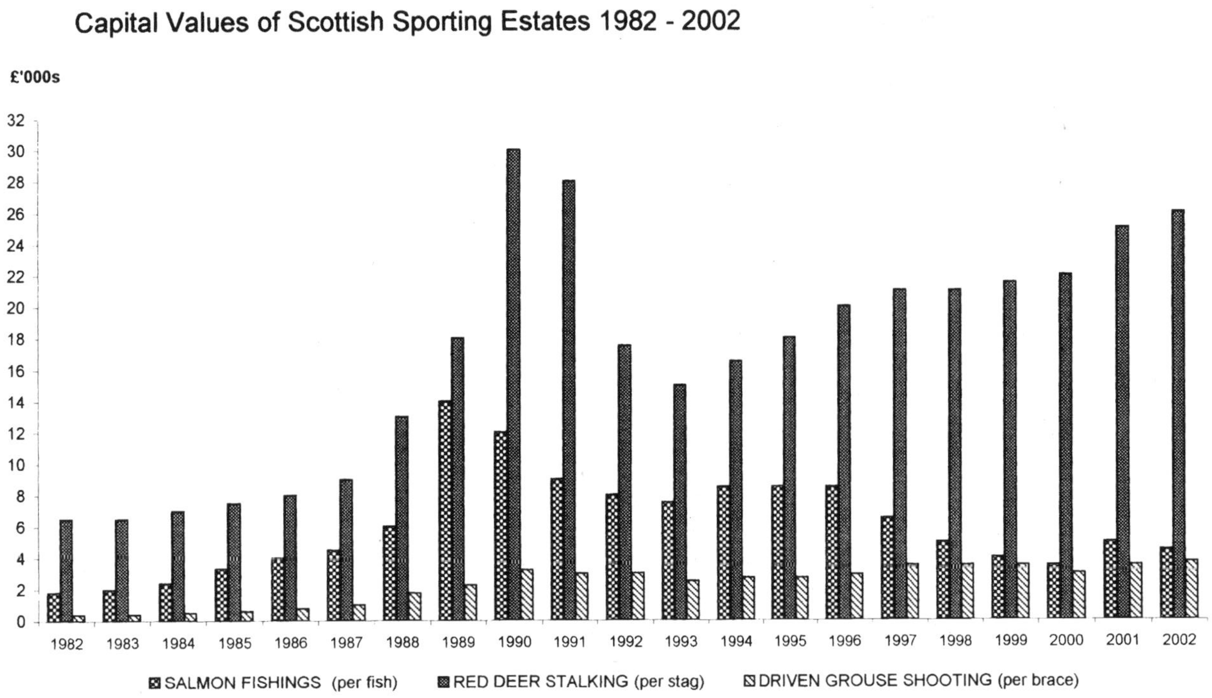

Figure 11.1

Source: Strutt & Parker

11.4 Market trends

Figure 11.1 shows the approximate prices attributed to the three main highland sporting facilities since 1982. The fluctuations for salmon and stalking reflect largely the state of the economy at the time, not only in the UK but also throughout Europe and elsewhere, as Scotland attracts widespread international interest. The purchase of a sporting estate, or even renting the facilities on it, is a luxury and the prices paid are vulnerable to any financial downturn. To a lesser degree prices may also depend on the success or otherwise of the preceding seasons, as shown in the records. There have been years when, for a variety of reasons, salmon numbers have fallen or when grouse populations have declined. In recognition of such vagaries, valuations should be based upon an average assessment of numbers taken over a period of at least five or even ten years.

Check-list – Sporting Rights:

- The value of sporting rights, whether in themselves or as an enhancement of the property as a whole.
- Capitalization of potential rent receivable from sporting lettings.
- Capital values of sporting properties determined per head of game 'bagged' in previous years.
- Influence of location and the quality of the terrain.
- Concept of stocking and of wild game.
- Market forces and taking 5–10 year averages.

CHAPTER 12

Minerals

The identification and assessment of mineral deposits is a highly technical matter that can really only be undertaken by specialists, but there are nevertheless some crucial points that will need to be considered by every valuer of agricultural land.

12.1 Mineral Rights

The right to extract minerals does not always run with the ownership of the land, as some, such as oil and coal, are reserved to the State and others may well have been retained by a previous owner or acquired by a potential operator. Whether a particular property includes these rights or not will have to be established by reference to the title deeds. These may then also detail the terms under which the owner of the rights may extract the minerals and the level of compensation, if any, that would then become due to the landowner.

The impact on land value will depend upon the particular circumstances. If the rights are not reserved elsewhere and there is some reason to expect that the property does include workable and saleable deposits, then this would be likely to enhance the overall value. Similarly, if those rights were reserved away from the property but subject to a modern rate of compensation then again there would be a positive effect on value. It is, however, also possible that the minerals may be extracted with only negligible compensation or reinstatement in which case the property might be blighted by the damage that such an operation would cause.

Having established the legal position, one needs then to find out whether there is any likelihood of commercial deposits of minerals under the land and whether they could ever be extracted.

12.2 Deposits

In the first instance deposits may be identified by reference to the Minerals Local Plan held by the County Planning Office. Such plans will indicate not only where potential deposits have been identified, often by minerals companies preparing for possible future workings, but will also show areas of strategic reserves. In the latter case it is possible that the land could be acquired under compulsory purchase orders at a later date as and when those particular resources might be needed. Depending on where such deposits lie within the property and whether they form part of a larger reserve, the value of the land itself may well be affected. If the deposits form part of a more general record, the potential effect on value will depend on the scale and quality of the deposits and on their accessibility and the likelihood of planning consent ever being granted.

The scale and quality of the deposits can really only be determined by reference to geological surveys that will have had to be carried out on the site, although the fact that these areas have been identified on a Local Plan would suggest that some initial investigations have already been made and have established at least the possibility of commercial workings. Whether such workings would be prejudiced by problems of access or by local planning considerations can be more readily assessed initially by a general analysis of the local situation. For example, it may well be more difficult to gain planning consent for mineral extraction from sites that lie within an Area of Outstanding Natural Beauty than for those in less sensitive locations. Similar difficulties may also arise where the site is so positioned that it can only be reached by passing through residential areas or along narrow lanes. Much may depend too on the type of aggregate being extracted, the local needs for such material and the anticipated length of time and scale of operations involved.

The future reinstatement of the site will have a bearing on these issues as well and should therefore be considered by the valuer as part of the overall assessment. In areas close to towns, there may be a greater pressure for landfill sites such as would be created when minerals had been extracted and this may influence the planning authority. In other locations it might be swayed by the potential for improving the immediate environment in the future through landscaping and water-based recreation.

12.3 Valuation Factors

There are therefore some general factors to be considered wherever the ownership of land is affected by the existence of mineral deposits.

First, it needs to be established whether the landowner is likely to benefit or suffer from the exploitation of the minerals. In the latter case, which could arise under one of the more traditional clauses whereby the rights are reserved away from the property with provision for only nominal compensation, it will be a question of the amount by which the land is then blighted. This will in turn depend on the potential scale of any extraction, the likelihood of it taking place within the foreseeable future and, of course, the impact that it would then have on the viability or enjoyment of the property.

Even where the owner stands to gain financially from a mineral operation, these factors will still have to be taken into account. The returns secured from a well negotiated agreement on royalties and reinstatement can be substantial but there may also be considerable disadvantages arising from the operation too. These may be short term, such as loss of farming or shooting income during the period of extraction and reinstatement, or more permanent as in the case of land that can no longer be used for building development or may not have been reinstated to full agricultural use.

These various factors can be relatively easily identified, provided that one takes a suitably comprehensive approach to the situation, but it is more difficult then to find a means of quantifying them within the valuation. First, there is unlikely to be any evidence of actual market transactions. Mineral-bearing land may occasionally be bought by operating companies and held in a land bank for future exploitation, but it is more usual for this to be done under an option agreement or conditional contract with the landowner rather than as an outright purchaser.

Even if there is evidence of such land being acquired, it may not be publicly known what quantity or quality of minerals it might contain. Local circumstances will of course also vary hugely according to the likely demand for material. Aggregates are heavy and thereby expensive to transport and so there will be a premium on sites that are accessible to forthcoming developments such as housing or road construction. Correspondingly, there will also be a discount on those in more remote locations. The depth to which the deposits lie below the surface will also have a bearing on the cost of extraction and thus on site value too. Other such costs can arise from conditions attached to the granting of planning consent particularly for reinstatement.

Essentially the impact that mineral deposits may have on the value of an agricultural property could be measured as a capitalization of the income received through royalties, as this would be the benefit that a landowner would normally seek to gain rather than making an outright sale of the freehold of the land. Such calculations may need to allow for the fact that mineral extraction is often done in phases over a long period of time. Much of the income will therefore have to be on a deferred basis, effectively as a Discounted Cash Flow. It is likely too that the agreement will include an arrangement whereby the operators are protected to some extent against a fall in demand. They will then be committed to working only a specified minimum level of extraction, and thereby of royalty, so that the income received by the landowner would fall for that period. This will need to be identified and incorporated within the assessment of the level of future income flow and the degree of risk attached to it. Such an analysis, although specific to the nature of mineral leases, would follow the general valuation principles outlined in Chapter 5.

It would however be unusual to rely solely upon an investment method to derive a market value in the way that one might do in the case of tenanted farmland. In contrast to the assessment of a complete farm or estate, the minerals may only affect one part of a property and also, of course, there are often many more mitigating factors involved as just referred to previously.

Check-list – Minerals:

- Check the ownership of mineral rights and the possibility of deposits.
- Value of minerals dependent on quality, accessibility, location and planning constraints.
- Value to landowner dependent also on terms of contract with operator.

CHAPTER 13

Compulsory Purchase

In principle, land that is being taken for some statutory purpose under a compulsory purchase order should be valued as if it were being sold in an open-market transaction. In practice, however, such cases will often be combined with wider negotiations that will need at least to have been noted by the valuer.

If the property being acquired happened to be a separate parcel of in-hand land that could be readily disposed of without any particular effect on other property being retained by the owner, the purchase figure would be assessed exactly along the lines defined for Open Market Value, as referred to in Chapters 1 and 2. In practice, however, the sale of such land may involve the vendor in related losses or inconvenience that will need to be compensated and therefore quantified. Although such payments or remedial works will be additional to the agreed value of the land itself, they should still be recognized and allowed for when deriving the appropriate market figure.

This may be illustrated by the example of a new road being constructed and passing across a farm. The area of land being taken up by the road itself would be valued by reference to all the normal market factors described in previous chapters. Generally this may be for a few acres of bare land for which there is likely to be local market evidence as single fields or 'accommodation land' is sold. Care will of course have to be taken in allowing for the effect of purchasers with a special interest in such cases, as with bids from neighbouring landowners, for not only would this lie outside the Red Book definition of Open Market Value (Section 1.3) but it would also not be appropriate to a strip of land running through a farm.

If the land is tenanted, then again the nearest market equivalent would be used, making the necessary allowance for the terms of the lease agreement and the various other factors mentioned in Chapter 4. In this instance, however, the fact that the site forms only part of a farm has different implications. As opposed to the vacant market where small parcels of bare land are traded quite frequently and often at premium prices, there is rarely any evidence of the equivalent occurring with let land. If such smaller plots of bare let land were to be sold, it would be unlikely to attract the normal investment buyer and could well in fact go to the sitting tenant who would of course be considered as a purchaser with a special interest.

There will be occasions when working out the equivalent market price of just the land being taken will not fully cover the loss in value to the owner. It may be that the property that remains will then be worth less per hectare than before if, for example, the house is going to be affected by noise from the new road or if, due to its reduced size, the remaining unit is no longer as economically viable as before. This situation, which is known as injurious affection will be the subject of a separate calculation from that for the purchase of the land itself, but will often involve valuation principles. Assessing the effect of disturbance from noise will be mainly a matter of market evidence and interpretation, whereas the question of viability may involve additionally a financial analysis and farm budget.

In practice, some of these features will be compensated by physical remedies or 'accommodation works'. For instance, the effect of noise may be reduced by the Highways Authority agreeing to build embankments or planting screens or by paying for double-glazing. The loss of land due to severance might be overcome by constructing a bridge between the two remaining parts of the farm or by providing an alternative plot taken from a neighbour. Such accommodation works are often of a practical nature and are agreed through negotiations that are not enforced by legislation and are not directly based on questions of value. These issues will therefore lie outside the remit of the property valuation, although ultimately whenever land is taken under compulsory powers, legislation requires that the owner is fully recompensed for any diminution in value that arises from it. Any disturbance or loss of income that has not been remedied by practical means will therefore need to be reflected in the assessment of value.

A public works scheme might be expected normally to result in a reduction in value or 'blight' to the property on which it is taking place. It can occur, however, that there is instead some improvement in value, or 'betterment', resulting perhaps from improved road access or revised zoning on the Structure Plan. The valuer will need to be alert to either of these possibilities when assessing the full implications of a proposed scheme.

The *Land Compensation Acts* were devised to determine the principles whereby statutory authorities are to ensure that landowners are properly paid for land taken and do not, as far as is possible, suffer any diminution in value as a result of such public works. In recent times, many of the public functions have been taken over by private companies. There is therefore a view that compensation being paid by these new utilities should include an element for the profit that they will be making from the work being done and should not be confined simply to compensating the landowner for any loss. Whilst one should be aware of this debate, it does not as yet affect the valuation principles outlined above.

In cases involving Compulsory Purchase the valuation of the property should follow exactly the same principles as for any other market situation. It will not in itself be influenced by the wider issues of losses and compensation, although an understanding of such matters is useful. Detailed explanations of how statutory compensation is applied to the agricultural sector may be found in *Agricultural Valuations – A Practical Guide* by R. G. Williams.

Check-list – Compulsory Purchase:

- Valuation as for open market, but with supplementary negotiations for compensation for loss and disturbance.

CHAPTER 14

Insurance Valuations

The valuation of property for insurance purposes follows different principles to those used in establishing a market value or realization price. It is in fact essentially a case of determining the costs that would be incurred if the premises had to be rebuilt or replaced and is not therefore influenced by the various property market factors that have had to be considered for all the other forms of valuation.

Providing estimates for building work is a specialized and technical matter, however, which can frequently be simplified in this context by taking overall measurements of the buildings and applying an appropriate cost per unit for the relevant type of construction. For the more straightforward structures, these figures can be obtained from a published source, such as those noted below, but there are then a number of further considerations that will still need to be taken into account.

In practice, building costs are likely to vary from the standard figures depending on differing circumstances which a valuer will need to be able to identify. These may involve, for example, the location and accessibility of the site, the scale of the work, and whether there are special features that would create extra expense or allow for savings. Most of these factors will be readily recognized and anticipated by the general surveyor, although the actual amounts to apply may need a more specialized and local knowledge. It will also be necessary to allow for the particular conditions that might arise following an insurable loss such as fire or storm damage.

One such allowance will be for costs which would not be included in the standard figures for new construction, but which are likely to be incurred in preparing a site for a replacement building such as the demolition and removal of the remains of the former structure. It is customary also to include an item for architects' or surveyors' fees for overseeing the work.

Replacement work of this kind currently attracts VAT, but as most farm businesses are registered for VAT and therefore able to reclaim this tax, insurance valuations are generally given as a net figure so as not to inflate the overall amount unnecessarily and thereby attract a higher premium. For private houses, on the other hand, the owners would not normally be able to reclaim VAT and the valuation would be grossed up by the appropriate prevailing amount.

There might be a saving also in allowing for an old property to be replaced with a more modern structure. This would not only be cheaper than replicating the original in traditional materials, but could additionally be more convenient to the occupier. This may be in entirety, such as replacing an old brick and slate tile barn with a portal frame clad in steel sheeting, or in part, as in the case of a stone building being reconstructed in concrete blockwork but clad externally in the original stone. Whether such measures are to be adopted by the valuer will depend on the client's instructions and on the feasibility of making such modernized alterations and, above all, on whether the original building is listed as being of special architectural or historical interest. In this last case, it would strictly speaking be necessary to estimate for the cost of rebuilding the property in traditional materials, which can be quite disproportionate to the modern usable worth of such a building.

Whether full insurance cover is then taken out is of course a matter of policy and is beyond the scope of a valuation, which is concerned purely with providing the appropriate rebuilding costs incorporating the relevant physical and monetary factors.

Check-list – Insurance Valuations:

- Determine cost of rebuilding by reference to standard unit costs.
- Make appropriate adjustments for local circumstances and for scale.
- Allow for site clearance, fees and VAT where relevant.
- Adjust for use of cheaper modern materials.

Sources for building costs:

General:

Spon's Architects' and Builders' Price Book.
Laxton's Building Price Book.

More specifically for farm properties:

Farm Building Cost Guide, from Buildings Design Group, SAC, Craibstone, Aberdeen.
Farm Management Pocketbook by John Nix from Andersons, Melton Mowbray, Leicestershire.

CHAPTER 15

Check-list for the Non-Agricultural Surveyor and Other Professionals

The following notes are intended as guidance for those who may become involved in a valuation that includes farmland but who do not themselves have agricultural knowledge. It may then be useful to be able to identify the main factors that need to be considered and to assess whether a specialist should be instructed and what features are then likely to be of particular significance. Such situations may arise when a property which is essentially of a commercial or residential nature includes an area of farmland, or where instructions to value a farm or estate are being given by someone working in a professional capacity such as a solicitor or accountant.

15.1 Small Areas of Land

Where a property comprises just a few fields of open land, the essential agricultural issues can be covered by the following basic questions.

i) *Is the land vacant or tenanted?*

If tenanted, is this on a statutory agreement or a modern Farm Business Tenancy? What are the terms of the lease and the age and circumstances of the tenant? What is the current rent and when would it be next due for review?

If vacant, is the land occupied by the owner himself or is it being farmed under some contracting or licence arrangement? In the latter case, is it legally certain that vacant possession can be obtained within a specified period of time?

As there are very few occasions when small parcels of tenanted land are sold in the open market, it can be difficult to find any valid comparable evidence. One needs then particularly to address the implications of a sitting tenant being a purchaser with a special interest or of using vacant land values and applying a percentage discount to arrive at an equivalent investment figure.

ii) *Is the land arable or pasture?*

If arable, has it been registered for IACS? Is it of a size to suit modern machinery? Would it require under-drainage and, if so, when was this last carried out?

If pasture, is the land unploughable and therefore without arable potential or was it previously cultivated and therefore registered for IACS? Is it fenced and provided with a water supply? Is milk quota

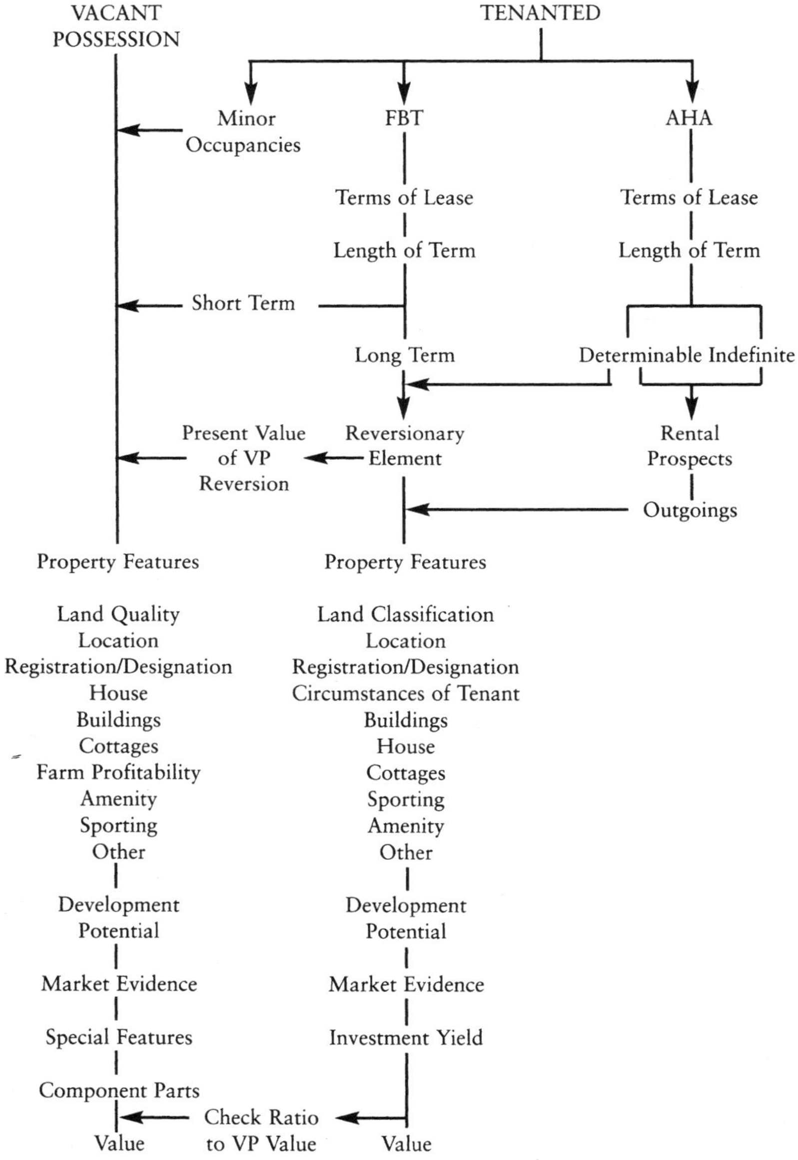

Figure 15.1 illustrates the basic process of appraising agricultural properties. It shows also diagrammatically the occasions when the valuation of tenanted land may need to be cross-referenced to a hypothetical figure for the property as if it were with vacant possession.

attached to it? Will the location attract demand, particularly for special interests such as horse grazing, or is it in an isolated position where stock might be at risk?

iii) *Under what grade, or class, has the land been classified?*

iv) *What indications might there be as to the quality of the land,* according to the type and condition of the crop, for example?

v) *Does it lie in a designated area* and, if so, does it qualify for particular grants or, conversely, are there special constraints or conditions imposed on it?

vi) *Can the land be divided into smaller lots* if this were appropriate to suit the local market?

15.2 Larger Areas and Complete Farms or Estates

Larger and more comprehensive properties will involve some further considerations in addition to those given above.

i) *Are the house and buildings suited to the farm* or are they outdated or representing undue maintenance costs? Are the buildings adequate to the modern needs of the farm? Are there cottages now surplus to these needs and if so how are they occupied?

ii) *Are any of the houses or cottages subject to an agricultural occupancy restriction?*

iii) *Is the farm connected to public water and electricity supplies* and is the latter adequate for equipment such as a grain drier?

iv) *Are there likely to be environmental problems*, particularly with regard to slurry?

v) *Is there shooting or fishing on the property* of significant value and, if so, what are the past records?

15.3 General

There will of course also be other factors that apply more generally to any property valuation and which will need to be considered in the context of farmland as well. These include questions of, for example, local market conditions, planning designations and access.

CHAPTER 16

Terms and Instructions

The terms under which a valuation is carried out and the content and purpose of the report need to be carefully defined so as to avoid any possible misunderstandings or disputes. All practitioners will have to satisfy themselves that the wording of their letters of instruction and of the report itself adequately fulfils this purpose and, where relevant, complies with the requirements of the Royal Institution of Chartered Surveyors. Such matters cannot be covered comprehensively in the context of an introductory book although an indication may be given of the main points that need to be considered.

16.1 Instructions

The letter accepting instructions should include the following:

16.1.1 The name and brief description of the property and the purpose of the valuation.

Although both client and valuer may think it obvious as to what property is being assessed, this brief description will ensure that there has been no misunderstanding about such matters as the extent of the ownership or whether any parts were to have been excluded from the valuation. The actual purpose of the valuation will have some bearing on the manner in which it is produced, as indicated in Section 1.4.

16.1.2 The source of information on the property and a note of the assumptions being made in addition to that information.

This effectively confirms the definition of the property and makes it clear as to what further investigations or measurements the valuer may have made. By stating the assumptions, one can establish also the limit of responsibility in certain areas. These are referred to below and will, for example, deal with structural surveys and planning enquiries. Sources of information will normally be from sale particulars or title plans provided by the client or their solicitors.

16.1.3 The Date

In a changing market, the date for which the valuation is to apply can be of crucial importance.

16.1.4 The Client

This may seem obvious in most cases where there is a single owner or purchaser requiring a valuation, but it still needs to be stated so as to limit the valuer's liabilities and avoid misunderstandings as indicated below.

16.1.5 Fees

As in all professional work, it is wise to agree fee arrangements in advance both as to the rate and to the time-scale for payment. With valuations, however, there is a further aspect in that an additional fee can be required if the report is to be used for securing finance and therefore passed on to a potential lender.

16.2 Terms, Conditions and Caveats

There are a number of definitions and qualifications that should be referred to in the valuation report. The nature and extent of these will be a matter of individual judgement, but are likely to include the following:

16.2.1 Structural Survey

A valuation appraisal rarely incorporates a structural survey, although it may well refer to one particularly where the condition of the property has a direct bearing on its market price. The report should therefore make clear that the valuer has not conducted a structural survey and should then define what inspections and assumptions have in fact been made. It may be mentioned, for example, that no inspection was made of inaccessible parts of the property and that throughout the report no comment is implied on the actual condition of the premises. This does not mean, however, that the property is being assessed as if it were in perfect order. The valuer will be making allowances for its perceived condition as mentioned in Sections 2.3.7 and 4.3.10, but will have done so without necessarily specifying the precise nature of the faults or becoming responsible for failing to identify such items.

Within this context, a valuer may also state specifically that no tests have been done on the ground conditions or on the services.

16.2.2 Contamination

The legal and valuation consequences of land or premises being contaminated are now so great that most reports will state that no investigations have been made on such possibilities and that the property is assumed to be free of contamination or hazardous substances. It may be added furthermore that the valuer, during the course of a normal inspection, found no evidence to suggest that an environmental report would be required. Where there may be cause to expect possible contamination, such as in the case of a farm incorporating an old landfill site, then the valuation report would be likely to include a recommendation that a separate environmental audit be carried out.

16.2.3 Planning

The potential or otherwise for development or, contrarily of blight, can be as important a feature as any of the property's visible characteristics. A valuer is likely therefore to mention what enquiries have been made of the relevant authorities, whether, for example, they were just a general verbal approach or a more specific written request. This will be determined according to the relevance of planning regulations and the likelihood or otherwise of development being permitted in the future or, indeed, of the value being reduced by the possibility of some adverse development in the vicinity.

16.2.4 Legal Interest

Unless there is some specific flaw or complication in the legal title to the property, the report will probably state that it is assumed that there is a good title and no unforeseen blight. This may include mention of the existence or otherwise of covenants, rights of way and maintenance obligations outside the property, such as in the case of a shared access. For a property described as being with vacant possession, it may still be mentioned that it is understood that no tenancies exist and in the case of a property that is wholly or even partly let there would of course also be mention of the presumed terms of the leases.

16.2.5 Tax and Expenses

An open-market valuation reflects as realistically as possible the price that would be paid for that property were it being bought and sold at the time. In actual practice both purchaser and vendor would face additional costs over and above this figure in the form of solicitors' and agents' fees and taxes such as stamp duty. The report may mention that no allowance has been made for such outgoings and that it has been assumed furthermore that the property is free of mortgages or other charges.

16.2.6 Disclosure to Third Parties

Due to the possibility of misunderstandings or misinterpretation, a valuation report will tend to state that it is confidential to the client to whom it is addressed and that no responsibility is accepted to any third parties to whom it may be shown. It is also generally stated that if the report is disclosed to anyone else, then reference must be made to the full terms and conditions under which it was prepared.

16.2.7 The Red Book

Reference is likely to be made to the valuation being conducted in line with the RICS Manual and the Institution's code of measuring practice, quoting possibly the relevant definitions from the Red Book. Such manuals do get updated from time to time and it would be appropriate to check that one is working from the latest edition.

16.3 Valuation Practice

Whether a valuation report is signed in a corporate name or by an individual valuer, it may include a brief description of the person responsible for preparing it and confirmation of their ability and qualification for this and that they have inspected the property.

It may be helpful also to add a commentary on the relevant market issues and to refer to specific examples and comparable cases that have been used. It is possible too to set out the workings by which the reported figure was achieved, although this is less common in the field of agricultural property valuations than in many other sectors. By quoting specific market evidence, one may be demonstrating the depth of research from which the conclusions have been derived but it will still need to be carefully qualified if one is to avoid it being misinterpreted by the client. Whether quoted in the report or not, all records of market evidence and of valuation workings should be kept on file in case they may need to be referred to in any future discussions.

It is worth repeating that the above comments cannot be a comprehensive and definitive model for a valuation report, but they should serve as an indication of the kind of approach that needs to be taken for reasons mainly of professional integrity. It may however be helpful to reproduce one example of a valuation report and that it is used for general illustrative purposes only. Therefore, in Appendix 1 will be found an extract from a sample report issued by the Agricultural Mortgage Corporation as a guide for use by the company's valuers. This is designed specifically for agricultural and diverse rural properties and sets out a useful, simplified structure for such valuations.

Check-list – Terms and Conditions:

- Ensure that the report fulfils the terms of the instruction, so as to eliminate scope for doubt or misunderstanding at a later stage.
- Define the task and the property and identifying areas not being investigated.
- Adhere to the Red Book.

Appendix 1

The specimen valuation report which is set out below is a format recommended by the Agricultural Mortgage Corporation for use by the company's valuers. It is reproduced here for illustrative purposes only and by kind permission of the AMC.

REPORT FOR VALUATION PURPOSES OF THE AGRICULTURAL PROPERTY KNOWN AS:	Grange farm, Ambridge, Loamshire
FOR THE PURPOSES OF A MORTGAGE TO:	P. Archer

---

1. SITUATION (location in relation to nearest towns/parishes)	Grange Farm is situated 5 miles from Penny Hassett and 12 miles from Borchester. The whole farm is contained within the parish of Ambridge as shown on the attached plan.
2. GRID REFERENCE	SP198074.
3. AREA	145.81 acres as per schedule attached
4. MAFF/MACAULAY LAND CLASSIFICATION GRADE	Grade 3

5. DESCRIPTION OF SECURITY (type of farm/brief description)	An extremely well-appointed dairy farm contained in a ring fence having most of the land level and productive pasture land. The farm is on or about the 600 ft contour line.
6. DWELLINGS (brief description)	There is a good 5-bedroom farm house recently constructed, a 2-bedroom bungalow, and a modernized 3-bedroom cottage.
(precise location)	The farmhouse is situated on the holding but the bungalow and cottage are situated in Ambridge village.
(construction)	The bungalow is a Woolaway construction, the other dwellings are of stone and slate.
(condition)	All the dwellings are in good order, although the farmhouse is in need of some decoration.
7. BUILDINGS AND FIXED EQUIPMENT (brief description) (adequacy)	The buildings comprise an extensive range of modern dairy buildings adequate for the herd of 100 cows.
(construction)	The majority are of steel framed and asbestos construction, a few older sheds being of stone and slate.
(condition)	The buildings are generally well maintained.
8. LAND (type)	The fields are generally well roaded. The soil is predominantly loam, although some of the lower land is heavier but still productive.

Pasture	135 acres
Woods	8 acres
Buildings	1 acre
Scrub	2 acres

(to be included in all reports)	The attached plan and schedule of land set out the area in detail and location of the dwellings and buildings.

(condition)	The land is exceedingly well farmed. Most of the fences, hedges, ditches and gates are in very good order.
9. TENURE	Assumed freehold.
10. TENANCIES (tenancies/occupation)	None a. The farmhouse is occupied by a Mr and Mrs P. Archer. b. One cottage is occupied by a son and daughter-in-law. c. The bungalow is occupied by a daughter. d. The shooting rights are let to Mr B Aldridge.
11. SERVICES/ABSTRACTION LICENCES	a. Electricity: mains electricity is connected to the dwellings and principal farm buildings b. Water: mains water is connected to all the dwellings, principal farm buildings and most of the land. c. Drainage: cesspool d. Abstraction Licence: provides spray irrigation from the River Am.
12. ACCESS TO PROPERTY (directions)	From the B3725 along a council adopted single concrete road, a quarter of a mile long.
13. MINES AND MINERALS	All rights are reserved but are not considered to be of any significant value. This has been taken into account in the valuation.
14. SPORTING RIGHTS	The woodlands are rented out for occasional shoots.
15. RESTRICTIVE AGREEMENTS	The bungalow is covered by an agricultural occupation clause. No other restrictions affect the property.
16. QUOTAS (all quotas)	A milk quota of 550,000 litres is attached to the holding, a copy of which is enclosed.

17. ENVIRONMENTAL FACTORS	Farm waste facilities / pollution. The farm waste facilities include an above ground slurry store. Yard water is drained to a collection pit which is connected to a low volume irrigator. No pollution evident. Contaminated land. Our enquiries have not revealed any contamination affecting the property or neighbouring property which would affect our valuation. However, should it be established subsequently that contamination exists at the property or on any neighbouring land, or that the premises have been or are being put to any contaminative use, this might reduce the values now reported.
18. RESTRICTIONS, RIGHTS OF WAY AND EASEMENTS (major points affecting valuation/details of essential access)	There are various rights of way over the farm but, apart from a well-used bridleway crossing OS100 as shown on the plan, no others affect the management of the farm.
19. OUTGOINGS	None disclosed.
20. TIMBER & WOODLANDS	The timber and woodland is of amenity value only.
21. DISCLAIMERS (by Valuer)	The 'special conditions' reference in the AMC covering letter to the borrower(s) refers to the appropriate disclaimer used by Agency firms.

22. VALUATION

I am of the opinion that the present open market value of this freehold property for mortgage purposes with the benefit of vacant possession, excluding quota, is fairly represented by the figure of £-.

I am of the opinion that the present open market value of the property, including the allocated milk quota of .... litres, is fairly represented by a figure of not less than £–.

Useful References and Further Reading

Agricultural Valuations – A Practical Guide. R. G. Williams. Estates Gazette. Manual on the valuation of agricultural business assets, including a chapter on farmland values and on rents.

An Introduction to Property Valuation. A. F. Millington. Estates Gazette. General background information to the essential principles of valuation.

Introduction to Property Valuation. David Richmond. Macmillan. Textbook covering essential principles of commercial property valuation.

Farmland Market. Edited by Louise Rose and Catherine Paice. Published bi-annually by *Farmers Weekly* in association with the RICS. A definitive source of statistical data on agricultural and woodland prices throughout Britain, including commentary on national trends and regional variations.

Financial Management for Farmers and Rural Managers. Martyn Warren. Blackwell Science. Providing guidance on the preparation of farm budgets.

Land and Estate Management. John Nix, Paul Hill, Nigel Williams & Jenny Bough. Packard Publishing. General background to the practical and legislative issues incurred in land ownership.

Laxton's Building Price Book. Information on current building costs.

Milk Quotas Explained. John Edwards. RICS Books. A guide to the background and present workings of the quota regime.

Modern Methods of Valuation. William Britton, Keith Davies & Tony Jackson. Estates Gazette. Introductory textbook covering a comprehensive range of valuation issues, including a chapter on agricultural property.

Parry's Valuation and Investment Tables. A. W. Davidson. Estates Gazette. Set of mathematical tables used in the calculation of investment values of tenanted property.

Practical Forestry for the Agent and Surveyor. Cyril Hart. Alan Sutton. Textbook covering all aspects of timber production and including information on valuations.

RICS Farm Price Survey. Published quarterly by the RICS from data collected and analysed by the Centre for Rural Studies at the Royal Agricultural College, Cirencester.

Spon's Architects' and Builders' Price Book. Taylor & Francis. Information on current building costs.

Statements of Valuation and Appraisal Practice and Guidance Notes. RICS Business Services Ltd. Manual setting out formal guidelines for property valuations, applicable as general principles to the agricultural sector, and including a section specifically on forestry.

The Valuation of Agricultural Tenancies: Art or Artifice. Charles Cowap. RICS. Assessment of methods for attributing value to statutory farm tenancies.

Valuation: Principles into Practice. Edited by W. H. Rees. Estates Gazette. General introductory textbook, including a chapter on agricultural property.

Glossary

Agricultural Occupancy. Refers to a residential dwelling that was erected originally under a planning condition that its use would be restricted only to persons engaged in agriculture, either on the farm on which it stands or elsewhere within the locality.

Aid Payments. Cover a range of subsidies and support payments made under the Common Agricultural Policy and in many cases dependent on the location or designation of the land.

Assured Shorthold Tenancy. Form of residential tenancy under the *Housing Act* 1988 and 1996 which can be terminated after the initial term (of not less than 6 months) and thereby ensure possession for the landlord, in contrast to lettings under the *Rent Acts*.

Blight. See **Planning Blight.**

Capitalization. The multiplying up of rental or other income to derive the investment value of an asset. (See also **Years' Purchase** and **Yield.**)

Classification. (See **Land Classification.**)

Comparables. Examples of properties similar to the one being valued and for which evidence is available as to the price achieved in a recent sale.

Compulsory Purchase. The acquisition of property by a statutory authority for a public works scheme under the *Land Compensation Acts* and requiring the establishment of market value as well as compensation.

Current Replacement Cost. The cost of purchasing the property in its present condition and function, including the benefit of planning consents.

Depreciated Replacement Cost (DRC). A means for establishing the hypothetical cost of replacing an existing building or other facility on a similar site, allowing for depreciation and obsolescence.

Designations. Areas of land specified for particular protection on either a voluntary or compulsory basis, with or without the payment of grants and subsidies, generally for environmental reasons, but also for historic sites and applying then to individual buildings as well as open sites.

Discounted Cash Flow (DCF). A calculation of the anticipated future income and expenditure arising from an investment property, used generally to establish a current capital worth of that net income flow.

Estimated Realisation Price (ERP). Market value assuming that the sale would be completed at a specified future date after having allowed a reasonable period for marketing, formally defined in the Red Book.

Estimated Rental Value (ERV). The anticipated level to which a rent would be reviewed as if that review were taking place under circumstances prevailing at the time in question.

Estimated Restricted Realisation Price (ERRP). Market value assuming that the sale would be completed at a specified future date but without allowing for a period of marketing, formally defined in the Red Book.

Existing Use Value (EUV). Incorporating the Red Book definition of Open Market Value but with the added proviso that for the foreseeable future the property be used only for its existing use. (See also **Current Replacement Cost.**)

Farm Business Tenancy (FBT). Tenancy created after the introduction of the *Agricultural Tenancies Act* 1995 and not therefore necessarily incorporating the statutory requirements and protections of previous *Agricultural Holdings Acts.*

Fee Simple. An absolute title to property in Scotland, equivalent to Freehold in England and Wales.

Forced Sale Price. A concept referred to in the context of mortgage valuations and equivalent to Estimated Restricted Realisation Price.

Freehold. An absolute title in property without obligation to pay rent or other dues to a superior owner. Sometimes mistakenly used to describe land with vacant possession.

In-hand. Implies that land is free from any tenancy or other form of statutory occupation by a third party and therefore being farmed by the owner, either directly or under a contract agreement. (See also **Vacant Possession.**)

Integrated Administration and Control System (IACS). A system of registration for all arable land within the European Union that was accepted as eligible for aid payments at the time of its introduction in 1991. Land not so registered at the time would not now be entitled to such payments.

Internal Rate of Return (IRR). A calculation of yield reflecting the performance of an investment allowing for all anticipated changes in income and capital positions during the term of tenure.

Land Classification. A means whereby all rural land in Britain is shown in map form as being of a particular class or grade. Intended originally as a tool for planning policy, it has also been used increasingly as a measure of quality for farmland and therefore influences value, especially for tenanted investment land.

Lease. General term for the letting of a property interest, ranging from agricultural tenancies to sporting rights.

Leasehold. Where the title in question is held under a long lease from a freeholder or head lessor. Rarely applied to rural property, but the term is often confused with that of tenanted land.

Licences. Either a short-term agreement to occupy land or a consent to a particular facility such as water extraction.

Lotting. The division of land into some of its component parts, with the intention of achieving a higher total sale price than if sold as a whole.

Model Clauses. Referring to the definition of landlords' and tenants' liabilities for maintenance and repairs under the *Agriculture Regulations* of 1948 and 1973.

Open Market Value (OMV). A basis of valuation, reflecting as far as possible normal market conditions and excluding certain special situations, formally defined in the Red Book.

Parry's Tables. A published set of tables enabling the calculation of investment values for a diverse range of income streams and forms of tenure. (See **Further Reading.**)

Perpetuity. Referring mostly to a source of investment income for which there is no structural reason to anticipate change, as in the case of a rack-rented statutory tenancy.

Planning Blight. The effect on market attitudes in a locality affected by building or other development, whether actual or proposed, and leading to a diminution in property values due to anticipated noise and disturbance, etc.

Premium. (See **Vacant Possession Premium.**)

Present Value of £1. Multiplier applied at an appropriate rate of return to derive the current worth of a capital item receivable at a specific date in the future, as in the case of a reversion to vacant possession.

Profit Rent. The difference between the rent being paid and either the income being earned by a tenant or alternatively the prevailing market level of rents.

Quota. A permit giving entitlement either to produce a particular commodity, such as milk, without payment of levy, or otherwise to produce stock and qualify for subsidy or grant assistance, as in the case of sheep or beef. In the former instance, quota is attached to land and has also had a tradable value of its own.

Rack Rent. The assumed full market rent at the particular time in question.

Recommended Sale Price. A figure reflecting sale tactics and not necessarily following the formal principles of valuation.

Red Book. A Manual produced by the Royal Institution of Chartered Surveyors and providing definitions of valuation principles and practice.

Reversion. An anticipated future change in circumstances in a property that is expected to bring additional returns or gains to the owner, as in the case of securing vacant possession or carrying out development.

Share Farming. An arrangement whereby an owner can allow his land to be farmed by another without the risk of creating a tenancy.

Shorthold Tenancy. (See **Assured Shorthold Tenancy.**)

Statutory Tenancy. In an agricultural context, implying a tenancy governed by the *Agricultural Holdings Acts* and conferring substantial security of tenure. Also, applied to residential lettings under the *Rent Acts*.

Succession. Implies the ability of someone qualifying under the terms of the *Agriculture (Miscellaneous Provisions) Act* 1976 to apply to become a statutory tenant in place of an earlier tenant who may have died or retired.

Tenancy. (See **Statutory** and **Farm Business Tenancy**.)

Vacant Possession. A status whereby the owner of a property has complete control over its occupation without the encumbrance of any ongoing tenancies or leases.

Vacant Possession Premium. The difference in value of the land when vacant and if subject to tenancy.

Worth. Calculation of Worth as a valuation concept defined by the RICS and referred to in Section 1.4.

Years' Purchase (YP). The multiplier applied to a rental income to derive its present capital worth. Also defined as the present value of one pound per annum receivable during the term of the tenancy.

Yield. The rate per cent return that the rental or other income represents against the value of the asset.

Yield Class. A measure of rate of growth of forest trees.

Zoning. A planning term specifying the range of development purposes for which an area so defined may be used.

Index

access 30, 153, 169
accommodation works 173
advertising 37–38
Agricultural Mortgage Corporation (AMC) 188
Agricultural Holdings Act
  of 1948 2, 67
  of 1986 3, 52, 67, 68, 77, 100, 102, 103
Agricultural Holdings (Scotland) Act 1991 3, 63, 69, 102, 103
Agricultural Land Price Index 40–46
Agricultural Land Tribunal 67
agricultural occupancy 34, 135, *see* Glossary
Agricultural Tenancies Act 1995 3, 52, 100, 102, 110, 115
Agricultural Valuations – A Practical Guide 174
Agriculture Act 1986 121
Agriculture (Maintenance, Repair and Fixed Equipment) Regulations 1973 63
Agriculture (Miscellaneous Provisions) Act 1976 68, 77
Aid Payments, arable 20, 122, 126
Allocated Quota 121
amenity 104
Arable Quotas 122
arbitration 100, 102, 103
Arbitration Act 1996 103
Areas of Outstanding Beauty (AONBs) 124
assignable leases 4
Assured Shorthold Tenancies 52, *see* Glossary
auction sales 47–48

bank base rates 56
barn conversion 25, 143
bed and breakfast 144
blight, planning 133–134, *see* Glossary
boreholes 31
budgets 97–100, 103
building regulations 140
buildings, farm 17–19, 25–26, 34–35, 74–75
Business Property Relief 149

Calculation of Worth 10, 11
capital gains 49, 149
capitalization 81–86, 106, 111, 113, 157, 169, *see* Glossary
cash flow 11
Childers v Anker case 1994 103
claims, succession 68
Common Agricultural Policy (CAP) 57, 118, 139
comparables *see* Glossary & market evidence
component parts, property 15–20, 22
compulsory purchase 10, 171–174, *see* Glossary

construction costs 18
contractors, farming 14
Control of Pollution Regulations 1991 30
cottages, farm 16, 22, 24
Country Life 37
covenants 137
Cowap, Charles 107
cropping, specialist 71–72
Crown Estates 55
current replacement cost *see* Glossary

Dairy Produce Quotas Regulations 1984 118
dairy units 26
deer stalking 161
deferments 88–90, 110
demand, potential 60
Department for Environment, Food and Rural Affairs (DEFRA) 40, 100
Depreciated Replacement Cost (DRC) 18, 33–35, 144, *see* Glossary
Designated Areas, designations 29, 123–125, *see* Glossary
developers, development 61, 77, 128–145
Development Value 129–133
Discounted Cash Flow (DCF) 92–93, 155, 169, *see* Glossary
discretionary trust 60
diversification 77–78, 139–145
drainage 31–32
Drainage Rates 32, 65
Duchy of Cornwall 55, 72
Duchy of Lancaster 55

easements 137, 138
Edwards, John 121
electricity 32
English Nature 124
Environment Agency 123
Environmentally Sensitive Areas (ESAs) 124, 125
equipment, fixed 17–19, 72–74
Estimated Realisation Price (ERP) 9, *see* Glossary
Estimated Restricted Realisation Price (ERRP) 9, *see* Glossary
European Community/Union 12, 118
Existing Use Value (EUV) 9, 145, *see* Glossary

farm buildings 17–19, *and see* equipment, fixed
Farm Business Tenancies (FBTs) 4, 68, 86, 102, 108, 110–116, *see* Glossary
Farm Management Handbook 96
Farm Management Pocketbook 35, 96, 97, 98, 99
Farm Woodland Premium Scheme 149
Farmers Guardian 37
Farmers Weekly 37, 121
Farmland Market 39, 40, 42, 100
Farmland Price Index – RICS 41, 43–46
fee simple 4, *see* Glossary
felling licences 149
The Field 37
Financial Management for Farmers and Rural Managers 96
fishing 162–165
footpaths 137
forced sale price *see* Glossary
Forest Mensuration 151
forestry 26–27, 147
Forestry Commission 148
Fountain Forestry 147
FPD Savills 59, 104, 147
freehold land 4, *see* Glossary

game keeping 157
General Permitted Development Order 1995 134
grain storage 17, 26, 72
gross margin analysis 98–100

Hart, Cyril 151
hedges 136
Hedgrows Regulations 1997 136

Hill Farm Allowances (HFAs) 124–125
Hope Value 133
houses 15, 24, 104
Housing Act 1998 52

illegal development 134–135
income 81
indicators, soil 23–24
inflation 91
in-hand farms 14, 76
injurious affection 173
Inland Revenue statistics 40
institutions
  financial 60, 62, 72
  traditional 55, 72
instructions 183–184
insurance 11, 175–177
  woodland 155
Integrated Administration and Control System (IACS) 20, 126, *see* Glossary
Internal Rates of Return (IRR) 93, 151, 155, *see* Glossary
internet 37
investment
  calculations 80–94
  market 55–61
  policies 61–76
  property 4–5
  scales of 62
  security 53, 87
  woodland 153
  yield 53–61
Investment Property Databank (IPD) 147
investors 6–7, 55–60, 153

journalists 38

key money 108

Land Classification 21, 69–71, 123, *see* Glossary
Land Compensation Acts 173
Landlord and Tenant Acts 52
leases 4, 63–65, 92, 114, *see* Glossary
Less Favoured Areas (LFAs) 124, 125
levies 119, 120
licences, water extraction 31, 123, *see* Glossary
life interests 90
Limited Partnerships 3
listed buildings 136
Livestock Quotas 121–122
local interests 19
location 27–29, 65
lotts, lotting 20, 48, 140, *see* Glossary

Macaulay Institute 22, 69
maintenance costs 30
market
  attitudes 53, 76
  change 91
  evidence & comparables 36–50, 100, 102, 147–148, 150, *see* Glossary
  features 2, 4–8, 15
  interpretation 42, 46
  seasonal nature 50
Market Value (MV) 9, 10
marriage value 6, 66
media 38
Michaelmas 50
milk quotas 104, 118–121
  market for 119–120
Milk Quotas Explained 121
minerals 166–170
Ministry of Agriculture, Fisheries & Food (MAFF) *see* DEFRA
model clauses 63, 74,84, *see* Glossary
money rate 89
mortgage 10

Net Current Replacement Cost (NCRC) 144
Nitrate Vulnerable Zones 125
Nix, John (*Farm Management Pocketbook*) 35, 96, 97, 98, 99

obsolescence 35
Open Market Value (OMV) 9, 10, 28, 50, 61, 172, *see* Glossary
Option Agreements 133

paddocks 138–139
Parry's Valuation & Investment Tables 85 90, 92, 93, 132, *see* Glossary
pension funds *see* institutions, financial
planning 129–137
  blight 133, *see* Glossary
  consents 29, 129, 130
  minerals local plans 167
plots, small 138
pollution control 18, 30
Potato Marketing Board 122
Practical Forestry for the Agent and Surveyor 151
prairie value 150
Present Value 88, 93
Present Value (PV) of £1 90, 93, 132, *see* Glossary
preservation orders 136
price guides 39–46, 49
Price Index, Agricultural Land – RICS 41, 43, 44, 45
Primelocation 37
private treaty sales 47, 48
productive capacity 103
profit rent 107, *see* Glossary
property
  components parts 15–21
  equipped 11
  management 62
  multiple/single units 62
  quality of 5, 53
  rates 88
  sales 38, 43, 46–50
Property Finder 37
purchase costs 33–35
purchasers 14, 52
pylons 138

quotas 104, 118–122, *see* Glossary

rack rent 82, 83, 84, 107, *see* Glossary
rates of return 81, 87
Recommended Sale Price 50, *see* Glossary
'Red Book' 8, 92, 102, 152, *see* Glossary
redundant buildings 74, 136
regional influences 6
Register of Sasines 49
reinstatement, site 168
relascope 152
Rent Acts 52
rent(s) 58, 76, 96–105
  relevant factors 103
  reviews 85, 86, 102–105
repairs 29, 30, 75
Replacement Cost 9
repossession 67, 77
Residual Method 131
reversion/reversionary value 66, 81, 87, 89, 93, *see* Glossary
  calculation 73, 87
rights of way 137
risk 89
roads 172
  also see access
roll-over 7, 8
Royal Institution of Chartered Surveyors (RICS) 8
  Farmland Price Index 41, 43–46
royalties, mineral 169, 170
Rural Payments Agency 119, 121

sales *see* property, sales
Scottish Agricultural Colleges 96
Scottish Farmer 37
Scotland 4, 22, 49, 52, 63, 69, 70, 102
seasonal market 50
security, investment 53, 87
services, water, etc. 30–32
Set-aside 126
share farming *see* Glossary
shooting 27, 157, 158–161

shorthold tenancy *see* Assured Shorthold Tenancies
sinking funds 92
Sites of Special Scientific Interest (SSSIs) 124, 125
sitting tenants 52, 61, 100, 114
Soil Survey 21–22
soil types 21–24, 69–71
sporting rights 156–165
Standard Quota 121
statistics, market 39–46
statutory tenancy *see* Glossary and tenancies
succession 68, 91, *see* Glossary
sugar beet 122

taxation 7, 10, 42, 84, 115, 149, 153–154
tenancies 14, 51–79, 95–108, *see* Glossary and Farm Business Tenancies
tenants, calibre of 66
tenant's fraction 121
tenure 2–4, 5, 14
termination 14
terms & instructions 182–187
timber consumption 154
Tir Mynydd Scheme 124
Town and Country Planning Act 1947 134
traditional institutions 55
Transfer of Undertakings (Protection of Employees) Regulations 1991 (TUPE) 33
top-slicing 89
trees, preservation 136

vacant possession 4, 5–6, 7, 13–35, 105, 111
  premium 105, *see* Glossary
valuation
  practice 11
  principles 8–10
  purpose 10–11
Valuation of Agricultural Tenancies 107
Valuation Office statistics 40
valuation report, specimen 188–191
Value Added Tax)VAT) 84
variable costs 97
voluntary designation 124

Wales, National Assembly 40, 100
Walton v Inland Revenue Commissioners 1994 & '96 107
Warren, Martyn 96
water supplies 31, 137
wayleaves 138, 153
Williams, R.G. 174
wind farms 32
Windthrow Hazard Classification 153
woodland 4, 27, 146–155
Woodland Grant Schemes 149
worth *see* Glossary

Years' Purchase (YP) 82, 86, 110, 157, *see* Glossary
Yield Classes 149–150, 152, *see* Glossary
yields on investment 81–94, *see* Glossary

zoning *see* Glossary